GENDER AND THE SUSTAINABLE DEVELOPMENT GOALS

This book sheds light on the important and mostly neglected role that gender plays in achieving the UN Sustainable Development Goals, doing so by investigating three key problem areas: empowerment, education, and infrastructure.

Starting with a theoretical and methodological framework, this edited collection contains 12 chapters from scholars and researchers from around the world. The book includes numerous case studies discussing the current status of gender equality relating to the SDGs. It reinforces the significance of gender for sustainable and just development, highlighting how women play a major role in work organization, disaster management, income, household maintenance, and mediation of knowledge. "Women" as a classification encompasses much diversity with many intersecting axes of difference; this book focuses on the excluded and disadvantaged majority social group, without imposing homogeneity on that categorization. Many chapters focus on critical situations occurring in the Global South, where these issues are highly prominent, and importantly, these contributions are written by local scholars. Finally, the volume provides pathways for basic and professional gender responsive education and innovation in the field.

The book will generate important discussions in interdisciplinary research and higher education settings focusing on sustainable development, gender, equality, human rights, and education.

Astrid Skjerven is Professor of Design Theory at the Department of Product Design, Oslo Metropolitan University, Norway.

Maureen Fordham is Professor of Gender and Disaster Resilience and is the Director of the Centre for Gender and Disaster at the Institute for Risk and Disaster Reduction at University College London, UK.

ROUTLEDGE/ISDRS SERIES IN SUSTAINABLE DEVELOPMENT RESEARCH

Series Editors: Peter Dobers (president@isdrs.org) and **Volker Mauerhofer** (vicepresident@isdrs.org)

The Routledge/ISDRS Series in Sustainable Development Research is the publishing platform of the wide community of 2500+ researchers engaged in multi-, inter- and transdisciplinary research on the need for and experiences with the transformation to a sustainable society. The International Sustainable Development Research Society aims to disseminate knowledge on sustainable development practices in all spheres of society which enables the acceleration of sustainability initiatives: cities, communities, production, innovation, consumption, biodiversity and ecosystem services, agriculture, climate, renewable energy.

This series welcomes submissions for high-level cutting-edge research books that push thinking about sustainability in new directions by challenging current conceptualizations and developing new ones. The series offers theoretical, methodological, and empirical advances that can be used by professionals and as supplementary reading in courses in sustainability science, environment and sustainability, development studies, planning, and a wide range of academic disciplines.

Good Governance and the Sustainable Development Goals in Southeast Asia
Edited by Rasyikah Md Khalid and Ainul Jaria Maidin

Gender and the Sustainable Development Goals
Infrastructure, Empowerment and Education
Edited by Astrid Skjerven and Maureen Fordham

For more information about this series, please visit: www.routledge.com/RoutledgeISDRS-Series-in-Sustainable-Development-Research/book-series/RISSD

GENDER AND THE SUSTAINABLE DEVELOPMENT GOALS

Infrastructure, Empowerment and Education

Edited by Astrid Skjerven and Maureen Fordham

Routledge
Taylor & Francis Group
LONDON AND NEW YORK

earthscan
from Routledge

Cover image: Nora Pincus Gjertsen

First published 2023
by Routledge
4 Park Square, Milton Park, Abingdon, Oxon OX14 4RN

and by Routledge
605 Third Avenue, New York, NY 10158

Routledge is an imprint of the Taylor & Francis Group, an informa business

British Library Cataloguing-in-Publication Data
A catalogue record for this book is available from the British Library

Library of Congress Cataloging-in-Publication Data
A catalog record has been requested for this book

ISBN: 978-1-032-00456-3 (hbk)
ISBN: 978-1-032-00452-5 (pbk)
ISBN: 978-1-003-17425-7 (ebk)

DOI: 10.4324/9781003174257

Typeset in Bembo
by Newgen Publishing UK

CONTENTS

CONTRIBUTORS

Idun Alman-Kaas Yr holds a BA in art and dissemination, as well as an MA in Art, Design and Drama – study specialization Art in Society from the Faculty of Technology, Art and Design at OsloMet-Oslo Metropolitan University. Idun is currently working for Christen Sveaas Art Collection as an artistic coordinator and registrar.

Venke Aure, Professor in art didactics at the Department of Art, Design and Drama, OsloMet – Oslo Metropoilitan University. Among her recent publications are: Aure, V., and Dahlsveen, H. (2018). The kiss: Utterance surrounding the utterance within. *Journal of Artistic Research*; Aure, V., and Gjærum, R.e G. (2016). Kjønn som Væren: Hvordan kan iscenesatte Kunstuttrykk bidra til Forståelse av Kjønnets Ekstsiens? *Drama: Nordisk Dramapedagogisk Tidsskrift*; and Aure, V. (2015). The Epistemology of the Gaze and Photographic Essays: A Basis for Understanding Visual Representations of a City. In: Aure, V. and Bergaust. K., eds. *Estetikk og Samfunn: Tekster mellom Samtidskunst og Kunstdidaktikk*. Fagbokforlaget.

Kristin Bergaust, Professor at the Department of Art, Design and Drama at OsloMet – Oslo Metropolitan University. She works as an artist, a researcher, and an educator. She is interested in transdisciplinary efforts to contribute to ecological and transcultural processes in urban contexts through artistic research methods and technological developments. She is head of the transdisciplinary research project FELT: Futures of Living Technologies. The following are among her recent publications: Bergaust, K. (2020). How to Set the Table for Collaborations: Artistic Sensibilities and Methods. *Acoustic Space*, and Bergaust, K. (2020). *Oslofjord Ecologies: Artistic Research on Environmental and Social Sustainability.*

Hilde Synnøve Blix, Professor in music education and aural skills at the UiT – The Arctic University of Norway, The Music Concervatory. Co-leader of the artistic research project "ArtVAPO–True North." Head of the research group "InOvation" – Critical art pedagogy. The following are her recent publications: Blix, H. S., and Waade, R. A. (2019). Soundpainting som musikalsk Læringsverktøy i Kulturskolen. *Journal for Research in Arts and Sports Education*, 3(1); and Blix, H. S., Mittner, L., and Gjærum, R. G. (2019). I kjønneste Orden? *Nordic Journal of Art and Research.*

Valeria Bucchetti, Professor at the Design Department of Politecnico di Milano, where she teaches "Visual Communication" in the Communication Design Degree and "Communication Design and Gender Culture" in the Design Master Degree (School of Design). She is Chair BSc + MSc in Communication Design. Her interest involves visual and gender identities in communication. She is a member of the board of the PhD Design Board and of "Centro di Ricerca interuniversitario Culture di genere." She won the "Compasso d'Oro" Design Award as the co-author of the project for the multimedia catalogue of the Poldi Pezzoli Museum (Milan, 1995). Her latest book is *Cattive immagini. Design della comunicazione, grammatiche e parità di genere* (2021).

Francesca Casnati, PhD candidate at the Department of Design, Politecnico di Milano, where she works with the DCxCG research group. Her research area concerns the intersection between the discipline of Communication Design and Studies. Since 2018, she is a Subject Expert for the course "Communication Design and Gender Cultures" within the Master of Science program of the School of Design at Politecnico. The following are her recent publications: Bucchetti V., and Casnati F. (2020). Icons: Normativity and Gender Inequalities. In: Bojanić S., Loddo O.G., Zubčić M., eds. *Phenomenology and Mind*, (17); and Bucchetti V., and Casnati F. (2019). The Education of Responsible Designers in a Male-oriented Society. In: *Gender and Women's Studies '19. Conference Proceedings.*

Boel Christensen-Scheel, Dean of the Department of Art and Craft, Oslo National Academy of the Arts. Her main field of research concerns art's relational and communicational capacities in an experiential and educational context. She has also worked with art's more explicit political content and ecological projects in an art educational context. Her most recent publications include the introduction to the anthology *Education for a New Museum*, which she also edited (2019), as well as several articles on user participation and audience development, amongst them Christensen-Scheel, B. (2018). An Art Museum in the Interest of Publicness. *International Journal of Lifelong Education.*

Bertha Cecilia Garcia Cienfuegos, Research Professor, National University of Tumbes. Among her recent publications are Strategic priorities for the conservation and management of forest genetic resources – RGF in Peru. Her recent publications also include Cienfuegos, B.C. (2017). Meeting of Forest Genetic Resources

Experts – Santa Cruz Bolivia. AECID Spanish Training Center, Bioversity INIA Spain and Cienfuegos, B.C.G. (2017). *Aprendizajes colectivos y reflexión sobre la práctica en el diálogo entre extensionistas rurales. Un estudio de caso múltiple realizado en Argentina, Colombia, Cuba, Ecuador, Guatemala y Perú*. CONICIT.

Maureen Fordham, Professor of Gender and Disaster Resilience. She is the Director of the Centre for Gender and Disaster at the Institute for Risk and Disaster Reduction at University College London. She has been researching hazards and disasters since 1988 and is an expert on gender and intersectionality; community-based disaster risk reduction; resilience, capacities, and vulnerability analysis, focusing particularly on the inclusion of a range of social groups, especially women and children. The following are her recent publications: Evertsen, Yadav and Fordham (Forthcoming, 2022). Gendered experiences of (im)mobility in the context of global environmental change, *International Journal of Disaster Risk Reduction*; R. Zehra Zaidi, Maureen Fordham (2021). The missing half of the Sendai framework: Gender and women in the implementation of global disaster risk reduction policy, *Progress in Disaster Science*, Volume 10; Yadav, P., Saville, N., Arjyal, A., Baral, S., Kostkova, P., and Fordham, M. (2021). A feminist vision for transformative change to disaster risk reduction policies and practices, *International Journal of Disaster Risk Reduction*, Volume 54.

Rikke Gürgens Gjærum, Vice rector, The Artic University, Harstadø. The following are her recent publications: Clancy, A., and Gjærum, R. G. (2019). Home as Place for Giving Birth: A circumpolar Study of the Experience of Mothers and Midwifes. *Health Care for Women International*; and Gjærum, R. G., and Ramsdal, G. H. (2015). Change the Game: Reflective Practice through Forum Theatre. In: Vettraino, Elinor Linds, Warren, ed.: *Playing in a House of Mirrors: Applied Theatre as reflective Practice*. Sense Publishers.

Martina Maria Keitsch, Professor in Sustainable Design at the Department of Design, Norwegian University of Science and Technology, Trondheim. Keitsch has worked for over 20 years in the sustainable development area, mainly with eco design, and -architecture, urban planning, stakeholder inclusion, ethics of sustainability and gender mainstreaming. Keitsch is teaching Bachelor, Master, and PhD students and has published several articles on the aforementioned topics. Her recent publications cover the following: Sustainability and Inclusion in Public Space. She has recently published he book *Transdisciplinarity for Sustainability: Aligning Diverse Practices* in cooperation with Walter J. V. Vermeulen, Walter (Routledge, 2020).

Miriam Kienesberger holds a master in Gender Studes and Political Science. She a Research Associate/ PhD Candidate at the Leibniz Institute of Ecological Urban and Regional Development (IOER). Her recent major research publication is: Wolfram, Marc and Miriam Kienesberger (submitted): Gender in sustainability transition studies: Concepts, blind spots and future orientations, in: Environmental Innovation and Societal Transitions (not yet published).

Elena Lazos-Chavero, Professor at Instituto de Investigaciones Sociales, Universidad Nacional Autónoma de México. Holds a PhD in Anthropology and Socio-Economics of Development, Ecole des Hautes Études en Sciences Sociales, Paris. Has had guest professorships in Zürich and other European universities. Has written and edited several books and more than 100 research articles and chapters. She has been awarded several honorary prizes. From the political ecology perspective, she has worked agrifood systems and territories, forest conservation, gender, power and culture. Her last publications: Lazos-Chavero, E. (coord.), 2020. Retos latinoamericanos en la lucha por los comunes. Historias por compartir. Ed. CLACSO, Buenos Aires, 221 pp; Lazos-Chavero, E. y T. Rivera (coord.). Horizontes y Experiencias Agroalimentarias en México: Entretejiendo actores, escalas y dinámicas de transformación. Vol 1 and 2. Ed. IIS-UNAM, México (in press); Lazos-Chavero, E., P. Meli, C. Bonfil, 2021 "Vulnerabilities and Threats to Natural Forest Regrowth: Land Tenure Reform, Land Markets, Pasturelands, Plantations, and Urbanization in Indigenous Communities in Mexico" Land Vol. 10, Issue 12, 23 pp, 1340; Lazos-Chavero, E. y M. Jiménez, 2022 "Vulnerabilidades rurales a partir del envejecimiento entre nahuas del sur de Veracruz" TRACE 81, CEMCA, pp. 132–161.

Hiroaki Matsuura, Provost and Vice President of Academic Affairs and Professor of Health Economics and Demogrpahy at Shoin University. He is an economist and demographer who is interested in the intersection between human rights and population health. His recent publication appears in American Journal of Public Health, Sexual and Reproductive Health Matters, and Population and Environment. He is currently an editor-in-chief of Biodemography and Social Biology.

Volker Mauerhofer, Professor and Chair in Environmental Science (Specialization in Social Science) at Mid Sweden University, Östersund. Currently, he is also a Professional Research Fellow at the University of Vienna and holding a Visiting Researcher position at the Environmental Law Centre of Meiji University (Tokyo/Japan) as well as a Distinguished Visiting Professorship at the National University of Malaysia (Bangi Campus). He is the co-author of over 120 peer-reviewed publications and held more than 180 presentations worldwide. Besides supervising students, he has executed academic and freelance research over 25 years in more than 55 countries.

Cloe Mirenda, PhD in Sustainability Sciences, Universidad Nacional Autónoma de México. She holds a Master's degree in Anthropology from Sapienza University of Rome. Her main research fields are social vulnerability, gender and climate change, political ecology. The following are among her recent publications: Mirenda C. (2020), "De la vulnerabilidad a la creación de capacidades frente a la crisis climática. Experiencia participativas en clave de género en la Costa Sur de Jalisco," in Poma A. y Gravante T., *Generando Conciencia sobre el cambio climático, Instituto de Investigaciones Sociales*, UNAM (2022). Mirenda C., and Lazos Chavero E., "Cultural vulnerability, risk reduction and gender equity: two coastal communities of Jalisco, México," *Environmental Hazards* (2021).

Lilli Mittner, PhD is a senior researcher at the Centre for Women's and Gender Research at UiT The Arctic University of Norway and leader of the Balance Hub at UiT. Lilli coordinated the project 'Gender Balance in Art Education' (2015–2018) and is currently leading work packages in the research and intervention projects 'Prestige. Gender Balance in Research Leadership' (2017–2022) and the ERASMUS+ project 'Voices of Women' (2022–2024). In her research Lilli brings aesthetic and feminist perspectives together. She has published on multimodality and gender, critical pedagogy and curriculum transformation, gender equality in higher arts education, and situated art intervention research. Together with Matthias Mittner she has recently developed the Balancinator, a web app to measure and monitor Gender Balance within organisations (https://shiny.uit.no/balancinator/).

İlknur Öner, Professor of Sociology at Fırat University, Elazığ. Her research interests include international comparative studies with field studies on disaster sociology, comparison of children's experiences as disaster victims, relocations of high dam migrants, gender issues, urban sociology, and the environment. She has published papers, book sections, encyclopedia sections, and book reviews related to her studies. In the recent two years, she has been focusing on her forthcoming two books on long-run follow-up studies about the experiences of high dam migrants, disaster sociology, experiences of Marmara earthquake child victims, and experiences of university students at Firat University following the Van earthquakes. She has just recently written an entry for Childhood Sociology Dictionary, Portugal.

Sarah Parsons, Managing Director of East West Interface. Senior Teaching Fellow, School of Finance and Management, SOAS, London. She has over 20 years of international business experience and runs a cross-cultural consultancy company called East West Interface. She has worked with many international businesses and governmental organizations to support them with their strategic understanding of cross-cultural business communications and how to become globally gender aware organizations. She has appeared on various media outlets (BBC World News, CNBC, BBC World Service) to speak about the differences in business communications and the gender roles. Her area of research is Gender in International Business specifically relating to the cultural differences between Japan and the West and has published articles in East Asia Forum, Women and Leadership Publication, and HR Agenda Japan.

Satya Maia Patchineelam, PhD candidate in Urban Governance and Development, Institute for Housing and Urban Development Studies of Erasmus University Rotterdam. She has taken part in several international research projects.

Astrid Skjerven, Professor of Design Theory at the Department of Product Design, Oslo Metropolitan University. She is currently the Head of the planning of a new interdisciplinary PhD program, Innovation for Sustainability, the leader of the working group Grøntme, and a member of the research group Design, Culture and Sustainability. Her special field of interest the relation between culture and sustainability and the impact of aesthetics on daily life. She has published widely in international scientific journals and has been the co-editor of two special issues on design in the journal Sustainable Development. She was the main editor of book *Design for a Sustainable Culture: Perspectives, Practices and Education* (Routledge, 2017), in cooperation with Janne Beate Reitan. She is currently the co-editor of the book *Nordic Design Discourses*, which is to be published by Bloomsbury in 2022, in cooperation with Rachel Gottlieb.

Maartje van Eerd, Assistant Professor in Housing and Urban Development Studies, Erasmus University, Leiden. Has taken part in several international research projects. The following are among her recent publications: Hossain Foishal, I. and van Eerd, M. (2019). Housing Transformation, Evidence from Bangladesh and India for Planned Resettlement. *Liveable City* and Van Eerd, M. and Banerjee, B. (2017). Yangon: Empowering Migrant Communities to Securing Housing.

Marc Wolfram is Professor Dr. and is Director of the Leibniz Institute of Ecological Urban and Regional Development (IOER) and Chair of Spatial Development and Transformation at Dresden University of Technology. Among his recent research publications are: Wolfram, Marc and Miriam Kienesberg: Gender in sustainability transition studies: Concepts, blind spots and future orientations, in: Environmental Innovation and Societal Transitions (not yet published), Wolfram, Marc; Borgström, S. & M. Farrelly, M. (2019): Urban transformative capacity: From concept to practice. Special Issue guest-editorial of: *AMBIO*, 48 (5), 437–448, doi.org/10.1007/s13280-019-01169-y, Wolfram, Marc (2018): Urban planning and transition management: Rationalities, instruments and dialectics. In: Frantzeskaki, N., Bach, M., Hölscher, K. Avelino, F. Eds. *Co-creating sustainable urban futures*. Cham: Springer. 103–125. doi.org/10.1007/978-3-319-69273-9_5, Wolfram, Marc; Frantzeskaki, N. & S. Maschmeyer (2017): Cities, systems and sustainability: Status and perspectives for urban transformation research. *Current Opinion in Environmental Sustainability*, 22, 18–25. doi.org/10.1016/j.cosust.2017.01.014.

1

INTRODUCTION

Astrid Skjerven, Maureen Fordham, and Martina Maria Keitsch

Gender and sustainability have been on the research agenda for decades, while perspectives and methods of research have undergone a development from political critiques of the male domination of women and nature towards a more complex, sophisticated, and inclusive analyses.[1] Still, there has been little systematic study of the coherence between the two themes.[2]

In the sustainability field, minor scientific attention has, for example, been paid to the different roles of men and women in everyday life, in decision, transformation and implementation processes, and in designing the social and physical environment. Correspondingly, the significance of women's contributions to sustainable development has been underestimated and marginalized. As, e.g., Johnsson-Latham (2006)[3] points out, in term of resource use women show more sustainable consumption patterns, while the lifestyles and consumer patterns of men are more resource-intensive and less sustainable than that of women. These patterns also mirror a gender difference regarding orientations towards the society and conceptions of self-identity.

The PISA report of 2006[4] found that when compared with boys, girls have higher levels of concern for the environment and a greater sense of responsibility for sustainable development starting as young as 15 (OECD, 2007, PISA 2006: Science Competencies for Tomorrow's World).

Competencies and skills of women are frequently regarded as less important than that of men by, mainly, male decision-makers. For example, in retrospective to a study from 1990, Levy (2010) points out that feminine styles of leadership lead to loss of authority in male-dominated economic environments.[5] Although great changes have taken place in the three following decades, it is still a fact that men manage societal affairs, politics, and governing, while women have the responsibility of household, upbringing, and care for younger and elder generations. In 2020, only 25 per cent of all national parliamentarians are women, compared to 11

DOI: 10.4324/9781003174257-1

per cent in 1995 (UN Women, 2021).[6] On the upside, women have also developed abilities of great value in the ongoing environmental crisis. However, if these abilities are underestimated and not utilized in socio-political settings, they cannot be integrated in a global approach towards sustainable well-being. Competence in care, empathy skills, and transdisciplinary learning abilities are particularly relevant to complement technical know-how and scientific theory regarding sustainable development, and according to a recent OECD study, girls already display these skills to higher degree than boys.

> Boys typically reported higher levels of skills in the domain of emotional regulation such as stress resistance, optimism and emotional control as well as higher skills in the domain of engaging with others such as sociability, assertiveness and energy. Girls in both age cohorts (10 and 15 years old), typically reported higher responsibility, empathy, co-operation, tolerance and achievement motivation.
>
> *(OECD, 2021)*[7]

Regarding political decision-making should be an urgent matter to include women in the development process at all levels of decision and government. There is a growing awareness of gender discrimination and its consequences in policy-making and business, and several approaches exists to mitigate inequities. The most prominent example is the United Nations Sustainable Development Goals (SDGs) established in 2017. The goals constitute a landmark in putting sustainable development on the global agenda and at the same time integrating gender issues in the sustainability debate. Gender equality is described as one of the main goals.[8] Goal 5 states: 'Gender equality is not only a fundamental human right, but a necessary foundation for a peaceful, prosperous and sustainable world'. All 17 goals are intertwined and interdependent, and the matter of gender constitutes a recurring and basic theme in all of them. The goals are a concretization of the main challenges that humanity is facing. Thus, they constitute an important basis for research- based action in the field.

The United Nations is an acclaimed, global organization with members from all parts of the world. It is a forum for consensus-seeking discussion and action, and it offers funding of peacekeeping and development aid. The organization's whereabouts are broadly covered in mass media. Consequently, the SDGs have received much attention and are on the agenda in politics, government, industry, and education. However, the UN reports show that only minor improvements have been made regarding, e.g., female political and managerial representation.[9]

The issue of gender and sustainability has at least two aspects: the different contributions of men and women to achieving sustainability, and the different impacts of sustainability on men's and women's lives.[10] This publication primarily addresses women, their living conditions, and how this affects the society as a whole. A main focus is on women in the Global South. Women in the Global South are the most affected regarding environmental impacts and socio-economic disadvantages and have the least opportunity to influence their circumstances through education or otherwise get out of a vicious circle of poverty and ignorance, facts that support the theory of intersectionality, coined by Kimberlé Williams Crenshaw in 1989.[11] According to Alvarez et al. (2013),[12] gender equality is both a human right and an indicator of sustainability, and unless women are encouraged to take part in the economic, social, political, and environmental arena, sustainable development will not be achieved.

In 1949, Simone de Beauvoir stated that 'you are not born a woman, you are made one'.[13] In 1990, Judith Butler argued that gender is a social construction.[14] In this book, the term gender is understood as the social attributes and opportunities associated with being male and female. They determine what is expected, allowed, and valued in a woman or a man in a given context. During the recent years, the complex definition on gender as such has become the main focus in discussions related to equality. In this way, the issue of women's conditions has become marginalized. In this book, we argue that women, i.e., those who perform the traditional roles as defined by the division of labour in home and in public life, are still unequally treated.

The approach in this issue is feministic, although pluralist and not based on any specific feminist theory. It is also practical and operational, with the SDGs as the pivot point.[15] It is meant to be a basis and inspiration source for further research and action on gender and sustainability.

Gender is increasingly acknowledged as a significant dimension of moving towards the SDGs (e.g., Leach, 2016),[16] and gender equality is a 'fundamental human right' and 'keystone of a prosperous, modern economy that provides sustainable inclusive growth' (OECD, 2017, p. 3[17]). The impetus of this book is hence to discuss complex global social and environmental issues by through a gender lens.

The contributions span a wide range of topics related to the SDGs. They also cover several continents and countries, and their different challenges: The Amazons, Peru, Fiji, Japan, and Turkey and Norway. Many of these are case studies written by women who are themselves subjects to the conditions in their countries, sometimes with reduced opportunity to higher education and scientific discourse. They represent voices to be listened carefully to and constitute a substantial part of the book. The publication also contains managerial, governmental, and educational approaches that contextualize the problems and point to conditions for and ways of improvement.

The publication has been divided into three parts, each representing key factors in the work with creating gender equality and sustainability: Part I Empowerment, Part II Education, and Part III: Infrastructure. Empowerment is understood as paramount means to achieve transition to a more equal and sustainable society. It requires major changes in governmental, legislative, and other power structures at various social levels and is therefore often met with decision resistance from the mainly male entities of power. Education is necessary to achieve a contextualization and awareness of own situation and means to transform it. In several countries, females are prevented from having adequate education, particularly at a higher level.

This means that they lack competence and opportunity to contribute to scientific discourse. Environmental and social infrastructure is paramount for creating a safe and environmental-friendly transportation and communication and is a matter of urgency in many countries in the Global South.

Part I: Empowerment starts with Miriam Kienesberger and Marc Wolfram's chapter *'Engendering Sustainability Transitions by Design'*. It is a thorough analysis of the discipline transition management, a highly relevant administrative tool to make societal change. Within the field of sustainability transition (ST) studies, gender still forms a largely neglected and rarely theorized tool of analysis and intervention. Drawing on feminist theories, the authors propose a critical analytical prism to explore the crucial role of gender in understanding social change and designing interventions conceived to prepare, initiate, and navigate sustainability innovations. By illustrating the influence that gendered structures, symbolic orders, and performances exert on phases, sites, and outcomes of TM, they display how ST studies as a whole could benefit from such a perspective. This step towards engendering TM should also further the consideration of gender equality and empowerment (SDG 5) as a central dimension and outcome of research and policies directed towards STs.

Volker Mauerhofer's contribution *'Gender within Sustainable Decisions: A 3-D Sustainability Perspective'* addresses decision-making related to sustainability. Within Goal 5, it is stated: 'Gender equality is not only a fundamental human right, but a necessary foundation for a peaceful, prosperous and sustainable world'. This contribution addresses the question where within the three dimensions of sustainable development, the environmental, social, and economic ones, gender could be located for sustainable decision-making. In terms of methods, the assessment is conceptionally based on 3D-Sustainability, a decision support method for preliminarily prioritizing different options of action all contributing to a sustainable development. 3D-Sustainability, therefore, offers a set of criteria and preconditions for developing non-fixed hierarchies of priorities. In this way, it identifies sufficiency, ecological equity (consisting of intra- and intergenerational equity), socio-effectiveness, and socio-efficiency as sustainable development criteria (of 3D-Sustainability) most relevant to reaching sustainable decisions with regard also to gender. Detailed definitions of these criteria that allow to quantitatively assess them are presented and discussed in relation to gender. The application of this concept is shown by means of several practical examples. Particular operationalizing tools for such sustainable decision-making are further described for addressing – also in bottom-up manner – organizations and institutions. The contribution additionally results in the presentation of the contexts between the different elements as well as criteria provided by that sustainable decision-making method for the gender policy.

The chapter 'Gender Mainstreaming in Occupational Health and Safety: Challenges in the Era of the Sustainable Development Goals' by Hiroaki Matsuura discusses the challenges and progress of gender mainstreaming (GM) in the field of occupational health and safety. Understanding the sources of gender differences in occupational health and advancing GM in this area can contribute to the achievement of Goal 3 (good health and well-being), Goal 5 (gender equality), Goal 8 (decent work and economic growth), and Goal 16 (peace, justice, and strong institutions) of the SDGs. In order to make the workplace accessible to both women and men with no discrimination, all stakeholders must take into account diversity among all workers related to gender differences. The SDGs are certainly not a quick fix to solve all occupational health and gender inequity problems, but a strong vehicle to create better, healthier, more transparent, and more gender-sensitive workplaces. Relatively little attention has been paid to GM in the area of occupational health, but more progress will be expected in this area in the next decade under the SDGs.

In their chapter 'Gender Inequity Related to Climate Risks: Cultural Vulnerabilities as Obstacles to Sustainability in Rural Communities of Mexico' Cloe Mirenda and Elena Lazos Chavero present a case study from Jalisco in rural Mexico. Here, the incidence and intensity of extreme climate phenomena have increased in recent years, and cultural vulnerabilities have led to heightened gendered vulnerabilities. The aim of this chapter is to incorporate a gender perspective in understanding how cultural vulnerabilities influence populations' vulnerability to climate disaster in two rural communities on Mexico's west coast in the state of Jalisco as well as their vulnerability to climate variability in a rural community on the east coast in the state ofVeracruz. The authors have analysed more than 100 interviews on how women's cultural, social, economic, and political vulnerabilities to extreme climate phenomena are magnified by gender inequity on a local level. It fills a substantial gap in the research on the connections between climate risks and disasters on one hand and cultural vulnerabilities based on gender inequity on the other. The authors focus SDG 5 Gender Equality, SDG 10 Reduced Inequality, and SDG 13 Climate Action, which are regarded as intrinsically linked.

Part II: Education starts with Rikke Gürgens Gjærum, Lilli Mittner, and Hilde S. Blix' contribution 'Sustainable Gender(E)Quality: Gender in Competence Assessment in the Arts'. This chapter provides new insights into the concept of quality in higher education and connect the discussion to the SDGs 4, 5, and 9. More specifically, the authors explore in what ways gender matters for the ways competence is assessed in higher arts education in Norway. It is based on qualitative study that comprised 24 committee reviews and five intimate conversations with experienced committee members who assesses applications for full-professorship positions. What the theories presented, and the examples indicate, is that there is a need for increased exploration on how academia can contribute to real transparent discussions on quality, gender, and competence when it comes to the aim of achieving the global goal of sustainable and democratic knowledge production. The study shows that to build a more sustainable infrastructure within higher arts education, more predictable assessment criteria, meta-discussions on practices of academic assessment, and formal training of (new) committee members are needed. Those pathways will ensure more openness regarding promotion processes towards more sustainable educational and gender equality development, infrastructure, and innovation in the field.

Martina Keitsch's chapter 'Sensitizing Nepalese Students for Gender Mainstreaming in Sustainable Planning and Design' represents a rare example of how the Global North and South can learn from each other, exemplified by an educational course. GM is increasingly recognized in architecture and design education. This goes especially for sustainable planning and design, which acknowledges interdependencies of, e.g., SDG 5 Gender Equality, SDG 4 Quality Education, SDG 10 Reduced Inequalities, and SDG 11 Sustainable Cities and Communities. The following chapter presents GM within the curriculum of the Master and PhD program 'Energy for Sustainable Social Development' (MSSED) at the Institute of Engineering (IOE), Tribhuvan University, Nepal. The chapter briefly introduces the background of GM in politics and higher engineering education in Nepal and Norway. An overview of the MSESSD curricula with a special focus on the course 'Stakeholder Inclusion: Gender Mainstreaming and Energy Management' is presented. The chapter also discusses results and reflections on GM in the course, illustrated by student works and fieldwork examples in context-specific cultural settings. It appraises the feasibility of NTNU GM approaches for IOE/MSESSD conditions and summarizes findings from integrating GM in a sustainable design and architecture curriculum.

Valeria Bucchetti and Francesca Casnati's contribution 'Communication Design to Foster Gender Equality: Research and Experimentation in the Field of Education' analyses the many shapes of gender divide, which is currently at the heart of international debate and in their own country Italy. SDG 5 aims to 'achieve gender equality and empower all women and girls'. The chapter addresses the goal from a communication design perspective. From a disciplinary viewpoint, the issue concerns women's representation in the media and the repercussions of reinforcing stereotypes that lead to stagnant gender roles (points 5.4 and 5.5). Mediascape in Italy still plays a crucial role in conveying stereotyped models that influence self-perception and the resulting personal and collective identity. The article focuses on ongoing research and experiments conducted by the research group 'DCxCG' which means Communication Design for Gender Cultures. The group's activities are at the intersection of communication design and gender studies, mainly offering a critical view of the forms of gender representation and supporting new communication models in the educational context, from a multidisciplinary perspective. Starting from the schematization of the 'short circuit' created when the media use stereotyped models to reach the target, amplifying and affirming the stereotypes themselves, this work shows where and in which ways communication design can intervene with this vicious circle, focusing on the educational dimension.

The final chapter, 'Responsible Tech-Futures: Female Artists as Semantically Sustainable Tech Developers' by Boel Christensen-Scheel, Kristin Bergaust, Venke Aure, and Idun Alman-Kaas discusses the prevalent belief in digitalization as a solution to environmental problems. It addresses SDGs 5, 7, and 9. The authors challenge the notions that the future is digital and that technology development will contribute to a more sustainable living, simplifying and improving routines and production systems. Technology development does, however, also imply a production culture, with different agendas, value systems, and incentives. Few of these are explicitly tied to the justification for the production itself. In art and technology development by female artists however, the value system and potential sustainability is one of the driving factors in the production, underlining the semantic and ethical dimensions of technology development. The authors argue that female artists connected to the bio-art and media art movements propose a 'feminist' or 'queer' take on technological development and innovation cultures, where sustainability plays a vital and has a more fundamental position in relation to technology and society. The term 'semantics' is introduced, or re-introduced, as an ability to work with meaning structures and degrees of importance in technology innovation systems. Semantics can be tied to the entire meaning production around the system. The authors suggest that female artists can alter the sustainability consciousness through such a meaning producing awareness in art technology projects.

Part III: Infrastructure starts with the case study 'Changing Traditions: The Case of Amazonian Riverine Women and Belo Monte Hydropower Dam' by Satya Patchineelam and Maartje van Eerd. The article displays that the consequences of resettlement due to development projects to marginalized communities are greater than the loss of material elements. Losing traditions and spatial identity is a reality amongst those affected by it. The imposition of western-style projects and plannings has resulted into traditional communities changing their lifestyle and losing their livelihood due to weak policy planning and standard ideas of quality of life that is not compatible to the riverine reality. Riverine women that resettled to Altamira have lost their traditional identity and shifted their entire life to living the majority of their days in the city, changing customs, becoming more dependent on men and isolating themselves by deepening their responsibilities in the household. By using ethnographic method, with walking along and participatory approach, deeper understanding of the consequences created by resettlement was observed in the daily lives of the riverine women.

Sarah Parson's article 'Re-imagining the Gender Roles within Japan to Achieve Sustainability of the Japanese Population and Economy' addresses demographic issues in Japan related to SDGs 3, 5, 8, and 10. Although Japan must achieve 'sustainable gender equality' in order to balance the effects of a declining labour force, recent Government initiatives aimed at bringing more women into the workforce and increasing the numbers of women in leadership roles failed to make any progress on SDG goal 5 and its key targets: 5.1, 5.4, and 5.5. The entrenched cultural gender norms have led to gender segregation within the workplace, a widening gender gap and insurmountable barriers to the goals mentioned earlier. This chapter defines what 'sustainable gender equality' means for Japan, why it is important discusses some of these norms citing recent case studies that highlight the barriers they cause. The author concludes that the only way to achieve 'sustainable gender diversity' is through a realignment of these deeply entrenched norms from a grassroots level, suggesting ways this could be done using the socialization role of Japanese preschools and a blueprint recently announced by the Singaporean Government. It offers suggestions to how Japan could do this in a culturally appropriate way with unique opportunities to rethink the country's gender trajectory and achieve sustainable gender equality.

Ilknur Ôner's chapter 'Gender and Sustainability in Ovacik and Hozak Cooperative Acrivities, Tunceli Province, Turkey' addresses SDG 5 by investigating often neglected points such as the importance of togetherness of the activity networks of men and women; the need for a combination of local, national, and international actors; and their interplays in existing or newly emerging social texture. The paper examines how two types of income-generating activities of the earliest two cooperatives place gender groups in Hozat and Ovacık townships of Tunceli province in Turkey. Moreover, how their intragroup dynamics and groups' relationships within their members and communities in their specific environments shape group operations, and how the participants of these activities respond to the processes of the activities. The findings display that women and men have different roles in both activities within the embedded interactions and responses with cultural, social, political, ecological, economical, and other settings. Therefore, sustainability and survival of these cooperatives and the promotion of their activities require close multidisciplinary scrutinization and multifaceted support. This can be done with the vision of understanding networks without segregation to strengthen necessary social relationship areas and lines. These are closely in line with several UN SDGs, locally accumulated and activated culture and policies.

By shedding light on three key factors in gender equality and sustainability, we hope that the chapters in this book will become a vital contribution to the gender-sustainability discourse.

Notes

1 Macgregor, Sherilyn: *Routledge Handbook of Gender & Environment*. London: Routledge, 2017.
2 Ruth Meinzen-Dick, Chiara Kovarik, and Agnes R. Quisumbing: 'Gender and Sustainability'. *Annual Review of Environmental Resources* 2014, vol. 39, 29–55.
3 Johnsson-Latham, G. (2006), 'Do Women Leave a Smaller Ecological Footprint than Men?', Swedish Ministry of Sustainable Development, Stockholm.
4 PISA 2006, www.oecd.org/unitedstates/39722597.pdf
5 Levy, P. *Industrial Organizational Psychology: Understanding the Workplace* (3rd ed.). New York: NY: Worth Publishers, 2010, p. 372.
6 UN Women, Accessed 10 December 2021, Facts and Figures: Women's Leadership and Political Participation | What We Do | UN Women.
7 OECD, Beyond Academic Learning: First Results from the Survey of Social and Emotional Skills, Accessed 10 December 2021, Beyond Academic Learning: First Results from the Survey of Social and Emotional Skills | en | OECD.
8 https://sdgs.un.org/goals
9 https://unstats.un.org/sdgs/report/2021/ and https://unstats.un.org/sdgs/gender-snapshot/2021/
10 Macgregor, Sherilyn: *Routledge Handbook of Gender & Environment*. London: Routledge, 2017.

11 Crenshaw, Kimberlé W.: *Demarginalising the Intersection of Race and Sex: A Black Feminist Critique of the Antidiscrimination Doctrine, Feminist Theory and Antiracist Politics*. Chicago:The University of Chicago Legal Forum, 1989.
12 Alvarez, Michelle Lopez. From unheard screams to powerful voices: A case study of Women's political empowerment in the Philippines. 12th National Convention on Statistics (NCS), EDSA Shangri-la Hotel, Mandaluyong City, October 1–2, 2013 (2013).
13 Beauvoir, Simone: *The Second Sex*. London:Vintage, 2011.
14 Butler, Judith: *Gender Trouble: Feminism and the Subversion of Identity*. London: Routledge, 1990.
15 Presentations of various feministic and theoretical approaches are found in Macgregor, Sherilyn: *Routledge Handbook of Gender & Environment*. London: Routledge, 2017.
16 Leach, M. (ed.). *Gender Equity and Sustainable Development*. London: Routledge, 2016.
17 OECD: *The Pursuit of Gender Equality: An Uphill Battle*. Paris: OECD, 2017.

PART I
Empowerment

2

ENGENDERING SUSTAINABILITY TRANSITIONS BY DESIGN

How to critically unpack gender in Transition Management

Miriam Kienesberger and Marc Wolfram

2.1 Introduction

The current planetary state is marked by human-induced overall ecological degradation and sustainability challenges. This implies an environmental crisis, including, e.g. climate change, loss of biodiversity, toxic pollution, land degradation, deforestation, and ocean acidification. In turn, this environmental crisis is linked to a social crisis characterized by institutional failure, exclusion and aggravating disparities regarding wealth and risks, as well as political crises at all scales, generated by the continued inability to develop effective responses (Cielemęcka and Daigle, 2019, p. 67f). Given the extent of this distress, current forms of intervention seem to provide more of a symptom treatment than a promising approach to overcome these predicaments. Since the aforementioned sustainability challenges are entrenched in the prevailing structures of societal orders (Wittmayer et al., 2018, p. 81), large-scale transitions of socio-technical systems (e.g. agriculture, transport, energy) are needed to address the root causes of persistent problems. This means radical structural change resulting from a coevolution of economic, cultural, technological, ecological, and institutional developments at different scales (Rotmans and Loorbach, 2009, p. 185).

But how can such deep and disruptive transformations be achieved? This is a core question addressed by the field of *Sustainability Transition* (ST) studies, concerned with understanding and orienting complex dynamics of socio-technical change (van den Bergh, Truffer and Kallis, 2011). Given the demand for solutions to this pressing problem, ST studies have rapidly grown to form a widely cited source of knowledge and inspiration in policy and practice (Loorbach et al., 2017, p. 600). Within this field, *Transition Management* (TM) has been developed as a governance approach for sustainability innovation since the early 2000s, aiming to utilize the findings of ST studies in order to inform and shape the processes of transformative

DOI: 10.4324/9781003174257-3

change. TM has become adopted especially in Western Europe, but also transferred to various contexts in the Global South (Loorbach, 2010; Frantzeskaki, Hölscher, Bach, et al., 2018).

Proponents of this approach are positive that it provides a promising way to facilitate transitions towards sustainable societies: through the creation of spaces for searching, learning, and experimenting, "transitions can be influenced, supported, and accelerated by playing into existing dynamics" (Wittmayer et al., 2018, p. 81). However, the approach has also been criticized from the outset for various deficits undermining its claims. This includes especially the lack of a deeper understanding of individuals and agency, as well as the neglect of social conflict, democratic legitimacy, and power, thus reflecting an ignorant or naïve conception of transition politics (Shove and Walker, 2007; Meadowcroft, 2009). Such calls have partly been heard, and the debate about TM's future development is still ongoing (cf. Jhagroe and Loorbach, 2015; Frantzeskaki, Hölscher, Wittmayer, et al., 2018, p. 10).

It is exactly at this point that the present contribution wants to intervene, as the debate about TM is missing a vital dimension of societal change, namely *gender*. Even the critical discussion of TM in social and political terms leaves gender unmentioned (cf. Shove and Walker, 2007; Hendriks, 2009; Meadowcroft, 2009; Kenis, Bono and Mathijs, 2016; Frantzeskaki, Hölscher, Wittmayer, et al., 2018; Hölscher, Avelino and Wittmayer, 2018). This blind spot does not come as a great surprise considering the neglect of this social category throughout the ST field: only a few contributions have explicitly addressed questions concerning gender, and hardly any of them provides a well-founded theoretical background for doing so. While some discuss gender in relation to transition governance, none has used a TM perspective so far (cf. Wolfram and Kienesberger, submitted).

In the following, we want to make this fundamental conceptual gap visible and illustrate through examples, why the attempt to develop an approach for *transformation by design* such as TM, but also ST studies in general, should consider gender as an effective tool for developing critical analysis and effective interventions and ultimately achieving transformative impacts. In doing so, we refer to feminist theories, arguing that gender constitutes a multi-dimensional social category consisting of material and symbolic aspects of existence, and being constantly produced and reproduced in the course of ongoing social practices. In fact, gender is built into the arrangement of social life: it can be understood as a vital factor in organizing the daily existence of individuals, communities, and entire societies. More precisely, gender functions as a basic principle for allocating duties, rights, rewards, and power. Accordingly, gender should not be underestimated regarding its significance for shaping societal transformations. Therefore, theories and approaches need to be "*gendered*" to analyse and navigate trajectories of societal change, since addressing STs without a thorough understanding of gender in the re-organization of society and politics will necessarily remain "insufficient, unjust and therefore unsustainable" (MacGregor, 2009, p. 124; Kronsell, 2013; Walby, 2020). With this in mind, our contribution hopes to further gender equality and empowerment in the course of transitions, thus promoting the realization of Sustainable Development Goal 5

(SDG 5), and to deepen the understanding of such processes with a view to gender as an explanatory variable concerning all SDGs.

2.2 Gender and TM: Conceptual background

Since TM recognizes that contemporary societies face persistent problems which are deeply embedded in societal structures, these very structures form its subject of analysis and intervention (Rotmans and Loorbach, 2009, p. 185). Therefore, it is here that the role of gender must be addressed from the outset (Haug, 2005, p. 279). By using gender as a heuristic device, feminist accounts of, e.g. ecology or technology have uncovered gender relations as a key factor in broader societal change. More precisely, they have probed how gendered relations of power organize all systems and interventions, as well as how they are implicitly part of outcomes – including environmental ones (Faulkner, 2000, p. 87; Johnson, 2010, p. 36; Arora-Jonsson, 2014, pp. 289, 296). Accordingly, these approaches offer fundamental insights into structures and relations co-shaping socio-technical systems and can contribute to the core objectives of the TM approach.

2.2.1 Gender as an analytical prism for societal change

Gender forms a critical lens directed at understanding the construction of difference. Hence, it is neither an essential attribute of individuals nor a constant in social life (Acker, 2004, p. 20; Wallach Scott, 2010, p. 9f). Rather, gender requires interrogating how "the naturalized (and therefore unquestioned) binary male-female differentiation" and its heteronormative character (Winker and Degele, 2011, p. 55) are (re)produced, maintained, and challenged in the first place, therefore contesting the notion that gender differences are "natural" or "given" (Hawkesworth, 2005, p. 145). As Sandra Harding (1986) suggested, there are essentially three distinct processes through which gendered social life is produced:

> [I]t is the result of assigning dualistic gender metaphors to various perceived dichotomies [...]; it is the consequence of appealing to these gender dualisms to organize social activity [...]; it is a form of socially constructed individual identity.
>
> *p. 17f*

Following this widely adopted analytical triad, gender relations are therefore (i) historically developed through societal structures, (ii) symbolically and discursively constructed, and (iii) performed in interactions (Degele, 2008: 17). We address each of these qualities as a reference for the following discussion of TM:

1. **Structure**: Gender plays an integral role in the organization of social relations, most prominently the division of labour (Harding, 1986, p. 17), as manifested in the female responsibility for "private" care work and the gendered segregation

of occupations. Put differently, gender functions as a socio-structural category assigning one's place in the social order and framing the individual scope of action. However, it is also itself embedded in the societal reproduction (Degele, 2008, pp. 63–66, 76).

2. **Symbolism**: Gender equally configures symbolic and discursive orders as well as cultural fabrications, i.e. societies tend to organize the world around gender(ed) meanings. Numerous representations of social life and the natural world are rooted in gender difference, and gender is even assigned to non-human entities (Harding, 1986, p. 17; Degele, 2008, p. 19). This suggests that symbolic orders and norms produce binary conceptions of masculinity and femininity, which convey particular male and female practices as appropriate or desirable (Scantlebury, 2014, p. 187).
3. **Identity**: While structures and symbolism provide for gendered rules and meanings of social (inter-)action, gender is also deeply interwoven with people's sense of identity and expressed through their social practices (Harding, 1986, p. 17). Subjects do not simply pertain to a certain gender – rather, gender is something they do in everyday situations. This means that gender roles and relations as well as the gender binary itself can be understood as a product of repetitive performative activities (West and Zimmerman, 1987; Butler, 1990, 2006).

This conceptualization of gender as structure, symbolism, and identity is useful for orientation and systematization, and to specify theoretically the type of questions that can or rather must be addressed in order to understand the differences gender makes in terms of sustainability and the transition (management) process. It should be kept in mind though, that while complementary in theory, the interactions between these three qualities of gender are seamless. To capture the reciprocal effects between them, multi-dimensional analyses that connect between the particular and its broader context are therefore required (Faulkner, 2000, p. 92; Winker and Degele, 2011, p. 51). This is also a common aspect within ST studies, referring to transitions as disruptive co-evolutionary change in *structures, cultures, and practices* (van den Bosch and Rotmans, 2008).

Nevertheless, one should keep in mind that there are other social categories that mark all three qualities and must therefore be included in analytical efforts. This means that gender phenomena cannot be adequately understood in isolation from other axes of power (McCall, 1992, p. 838; Kaijser and Kronsell, 2014, p. 419). To reflect the interconnectedness of different forms of oppression as well as privilege due to gender, race, class, sexuality, age, ability status, and also species, the term *intersectionality* (cf. Crenshaw, 1991) has been coined (Sauer, 2018, p. 86; Lloro-Bidart and Finewood, 2018, p. 145). Therefore, any sustainability-oriented approach interested in socio/eco/technological issues, considering gender also demands an intersectional perspective for both diagnosis and solution. Given the scope of this contribution, we remain focused on gender here though.

2.2.2 How to "manage" a transition

TM constitutes a normative and prescriptive approach for governance activities and socio-political action in the context of complex societal problems. By combining insights into the dynamics of transition processes with insights from governance theory it formulates guiding principles for fostering STs, in other words, for influencing the direction and pace of transitions of societal systems into a more sustainable direction (Rotmans and Loorbach, 2009, p. 185; Loorbach and Rotmans, 2010, p. 237; Jhagroe and Loorbach, 2015, p. 67). TM emphasizes the need for

> a long-term perspective to guide short-term development, the acknowledgment of uncertainties and surprise, the importance of networks and self-steering, and the necessity of creating space for innovation.
>
> *Loorbach, 2010, p. 178*

TM offers operational guidance on how to set up a co-creation process for change-oriented interventions. Such processes are interactive and participatory in nature – they are based on selecting particular actors (e.g. change agents, frontrunners, and policy entrepreneurs) and engage them in long-term strategic envisioning of sustainable imaginaries, network formation, experimentation, and continuous reflexive learning (Jhagroe and Loorbach, 2015, p. 67; Frantzeskaki, Hölscher, Wittmayer, et al., 2018, p. 3; Hölscher, Avelino and Wittmayer, 2018, p. 132f). Thereby, TM provides mental, social, and physical space to develop new ideas, a common language, ambitions, and new joint projects. Put differently, it takes on a coordination function that brings inputs as well as streams of solution searching together (Frantzeskaki, Hölscher, Wittmayer, et al., 2018, p. 3; Wittmayer et al., 2018, p. 81f). Ideally, TM results in a shared sense of direction, initiates impulses for local change, and empowers actors to collectively tackle sustainability challenges (Roorda et al., 2014, p. 12). The goal is to create a societal movement, which puts increasing pressure on the political and market arenas to ensure that the transition process is in line with overall sustainability goals (Hölscher, Avelino and Wittmayer, 2018, p. 132).

In practice, the TM cycle consists of four different **types of interventions**: (1) **Orienting** is about building analytical capacity for transitions. Therefore, analysing and positioning oneself vis-à-vis societal developments, challenges, and other actors is central. (2) **Agenda-setting** involves building networking capacity for transitions. It is about broadening discursive and actor networks, creating a shared sense of ownership as well as ambition for a sustainable future. Thereby, it helps actors to integrate this ambition with their own agendas and practices. (3) **Activating** is about sharpening the orientation and enhancement of action. It involves putting shared direction into action through setting up projects and learning from them. (4) **Reflecting** is the primary means of learning from different interventions and therefore ongoing. It entails fostering a culture of reflexivity (Roorda et al., 2014, p. 10f).

These basic forms of activities are reflected throughout the seven **phases** of the process:

1. **Setting the scene** involves the formation of the *transition team*, which sets up and drives the TM process. This group specifies the issue to be addressed and thinks about potential outcomes (Roorda et al., 2014, p. 16).
2. **Exploring local dynamics** is about unravelling complexity. Therefore, the transition team develops an analysis of the societal system concerned. It defines temporal, spatial, and thematic system boundaries, as well as characteristics, indicators, and relationships (Roorda et al., 2014, p. 20f). Based on this system analysis, relevant actors potentially contributing to the desired transition are identified and brought together in the *transition arena*. This arena constitutes an informal, small group of change agents from diverse backgrounds (business, government, research institutes, and civil society) carrying out the further TM process (Frantzeskaki, Hölscher, Wittmayer, et al., 2018, p. 2f; Wittmayer et al., 2018, pp. 88, 92, 96).
3. **Framing the challenge** is about creating a shared understanding of the underlying problems. In this, the former system analysis is used as a point of departure (Roorda et al., 2014, p. 24).
4. **Envisioning futures** aims at drafting a long-term *vision*, which can be understood as a coherent storyline that sketches a future perspective. While this vision functions as an anchor point for strategies and can guide actions (Roorda et al., 2014, pp. 11, 15, 27), the process of envisioning itself is just as important (Loorbach, 2010, p. 175).
5. **Reconnecting long term and short term** consults this vision and translates it into *transition pathways* and a *transition agenda*. While the former describe inspiring narratives of change that include goals and action ideas for realizing possible routes from the present towards the envisioned future, the latter functions as a summary document containing challenges, visionary images, pathways, and action ideas (Loorbach, 2010, p. 175; Roorda et al., 2014, pp. 15, 30).
6. **Engaging and anchoring** involves diverse activities to make the transition agenda publicly owned. Therefore, the protected setting of the transition arena is left behind. The transition agenda provides a starting point for a wider group of people, organizations, and initiatives to adopt ambitions for a sustainable future, and relates this to their own agendas and practices (Roorda et al., 2014, pp. 15, 35).
7. **Getting into action** refers to the initiation and adaptation of *transition experiments*, i.e. radical short-term actions in line with the transition agenda (Roorda et al., 2014, p. 15). These actions expand the process outward into broader networks. Successful experiments can be repeated (broadening) and scaled up from the micro- to the meso-level (Wittmayer et al., 2018, pp. 88, 98).

2.3 Engendering Transition Management

Considering the above juxtaposition of TM with the notion of gender as a heuristic for analysing societal change, we have made the linkages between both perspectives visible. In what follows, we will further unpack where and how TM could adequately account for the fundamental role of gender in all social formations, drawing on its conception as *structure, symbolism,* and *identity*. In particular, this would be required throughout the process of TM in the *questions* it poses, the *outcomes* it envisions, and the *sites* it takes place in.

2.3.1 Structure

Gender relations are constitutive for institutional arrangements, and thus the (political-economic) order within which sustainability governance takes place (Hofmeister, Katz and Mölders, 2012, pp. 63–66; Kronsell, 2017, p. 110). With the

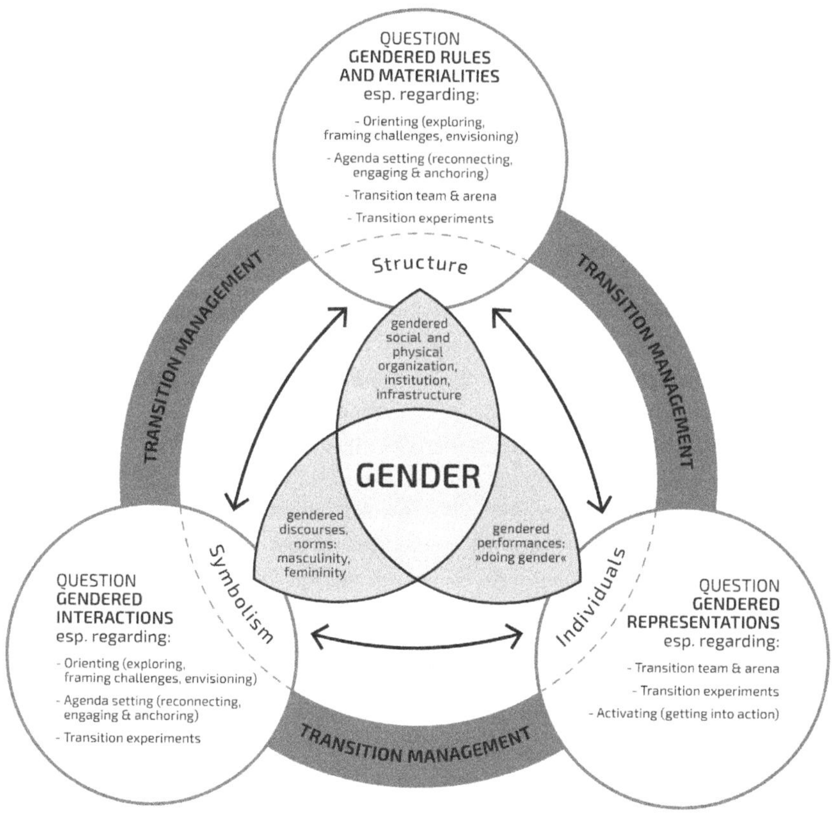

FIGURE 2.1 How to account for gender in TM (authors' illustration based on Paulus, 2015).

industrialization of capitalist societies in the Global North, social life has been increasingly divided into public and private spheres. This relates, for instance, to the separation between the commodity production in the public market economy and the unpaid reproduction in the private home (Acker, 2004, p. 23) and is reflected in material as well as in ideological terms.

This historical dichotomy of public and private is highly gendered and produces a particular structural form of masculine dominance: because the public sphere became associated with masculinity, while femininity was linked to the private sphere. Accordingly, men were assumed to be responsible for public life and economic activities, while women were regarded as fitted "by nature" for domestic reproductive tasks (Boyd, 1997, p. 8; Acker, 2004, p. 23). Over time, this gender-coded separation became a general principle of social organization and institutions, and continues to provide the grounds for the reproduction of different and unequal lives, as well as images and ideologies justifying them – until today (Acker, 2004, p. 24).

Therefore, TM's prospect to initiate structure-changing innovation dynamics seems limited as long as the gendering of the structures addressed is not made explicit. The whole TM process starts with and is based on the definition of an issue, the specification of intended outcomes, and a subsequent system analysis (cf. **setting the scene, exploring, framing challenges**). These initial steps are not just a technical matter but inherently political, since they involve construction as well as efforts of abstraction (Shove and Walker, 2007, p. 765). And as long as they are approached through a "patriarchal prism" – i.e. abstracting the public from the private and prioritizing the former over the latter – participants are blinded not only to deep operations of power relations, but also to possible practices and processes that may exceed and contradict binary logics (Youngs, 2000, p. 45f; Prügl, 2021, p. 296). For this reason, it is essential to pay particular attention, if feminized reproductive activities or inputs by the environment are recognized and addressed as integral part of, e.g. the food or energy system. Or if the focus solely lies on monetized processes associated with the masculine sphere.

A similar problematic applies to TM's methodology to formulate **visions**, **pathways**, and **agendas**, because these in turn refer to the challenge identified. Yet, taken-for-granted dichotomies embedded in the challenge are an untenable and impracticable prerequisite for imagining alternative futures (Prügl, 2021, p. 302). For example, if the role of social reproduction in the system was not identified, a market-based vision would narrow down the solutions considered plausible: e.g. economic instruments, productivity, technical innovation (Kronsell, 2013, p. 7f). Moreover, ignorance towards the domestic sphere and its gendered working prevents participants from grasping the effects presumed solutions can have on the private realm. For instance, that policies, placing responsibility for reducing environmentally harmful impacts on individuals, disproportionally burden women as carers and consumers (MacGregor, 2010, pp. 84, 93f) remains hidden. Therefore, an unchallenged gendered bias radically limits the range of conceivable futures and potential outcomes.

Furthermore, by considering the gendered structure of social life, questions concerning the composition and accessibility of TM's interaction settings (e.g. **team**, **arena**, **experiment**) arise since they can lead to an unequal distribution of power and influence (Hofmeister, Katz and Mölders, 2012, p. 66). Without such critical attention, it is no coincidence that a "masculinisation of environmentalism" can be observed. In other words, men dominate environmental decision-making at all levels – spokespeople, experts, entrepreneurs, and policy-makers (MacGregor, 2010, p. 84).

Therefore, not deliberately questioning the gendered logic of the socio-technical structures addressed has profound implications for TM and its impacts. During the orienting phase, it will strongly affect the perception and framing of the selected sustainability challenges and the vision pursued. These in turn determine to a large extent how they are addressed in the agenda and through experiments (Roorda et al., 2014, p. 24). A gendered dichotomous bias in the system analysis will necessarily be taken forward into the system innovation process.

2.3.2 Symbolism

Furthermore, gender constitutes a meaning system, which results in discursive and cultural constructions of masculinities and femininities shaping social life – more precisely, the ways people interpret, debate, articulate, and tackle socio/eco/technological phenomena. Masculinities/femininities produce, for instance, environmental politics and governance (MacGregor, 2009, p. 126f; Foster, 2017, p. 217). Therefore, becoming aware of how gendered discourses and norms find their way into the process of **framing challenges** and **formulating visions** appears particularly relevant.

In general, feminist approaches disclose an explicit connection between "masculine values" and dominant approaches to sustainability problems (Ourahmoune, Binninger and Robert, 2014, p. 324). By recurring to Cartesian thought, which binarizes and stratifies the mind over the body and technology over nature as well as associates the former with the masculine and the latter with the feminine, these discourses oftentimes echo traditional depictions of masculinity. Put differently, they are driven by rationality and based on the "modern" belief in progress, the glorification of techno-sciences as well as an anthropocentric view of the (hu)man as "master and possessor of nature" (Ourahmoune, Binninger and Robert, 2014, p. 319f; Foster, 2017, p. 221f).

For example, the discourse of ecological modernization which dominates international policies on sustainable development frames manifestations of the socio-ecological crises – such as climate change – as techno-scientific problems requiring "technical fixes" and claims that economic growth constitutes the basis for this post-carbon energy transition and a development towards sustainability (MacGregor, 2010, p. 90f; Hultman, 2017, p. 245). Therefore, these issues can apparently be solved by the partnering of two domains traditionally associated with masculinity: economics and technoscience. While recognizing the environmental problems and promoting an eco-friendly technology, this logic favours existing system solutions. It supports policies and technologies that – in the end – conserve the structures of destructive systems (MacGregor, 2009, p. 128; Foster, 2017, p. 226; Hultman, 2017, p. 246). And again, scenarios and visions developed along these lines hardly consider social agency, ignoring or trivializing, e.g. social reproduction as feminized sphere of activity (MacGregor, 2009, p. 131; Kronsell, 2013, p. 6). If framings and visions reflect discursive masculinity, this can have serious consequences for the transformativity and sustainability of the process.

In addition, it is necessary to pay attention to representations of men and women as well as their roles in TM outcomes: who is referred to and how – as private vs. public, vulnerable vs. independent, object vs. subject, subservient vs. dominant? For example, women are scarcely visible in techno-scientific discourses, while representations of males dominate. In turn, this does not apply to the feminizing discourse on green consumerism (MacGregor, 2010, p. 93; Foster, 2017, p. 226). Also, personifications as well as aesthetic codes used in narratives are of interest. For instance, is control and strength visualized by masculine bodies? Is nature constructed as symbolically feminine? Therefore, to promulgate symbolic gender norms in the TM process has at least two outcomes: it legitimizes certain environmental "solutions" and entrenches gender expectations (Ourahmoune, Binninger and Robert, 2014, p. 324; Foster, 2017, p. 216).

2.3.3 Doing gender while doing transition

Gender-aware perspectives encourage us to be interested in how those individuals participating in different TM settings (e.g. **team**, **arena**, **experiment**) *do* gender because it is constantly and unconsciously performed. Individuals construct feminine and masculine identities through repeated habits, thereby responding to context-specific social norms that prescribe how men and women should be and act (Shrestha, Joshi and Clément, 2019, p. 136; Dashper and Finkel, 2020, p. 73). For instance, the perception of modern high-tech as a male domain and "old" technologies related to cooking, laundry, etc. perceived as feminine can influence the ways men and women participate in processes, interact with certain technology as well as highlight or downplay their expertise in order to assert their gender identity. This can generate barriers for women to become fully engaged and can exclude them from being consulted – especially regarding financial and technical questions (Standal, Talevi and Westskog, 2020, pp. 4–8).

To obtain a differentiated picture, it seems vital to have a close look at the constitution of TM sites. Because gender is part of the occupational or organizational structures and forces certain kinds of gendered interactions, either a "doing of masculinity" or a "doing of femininity" (Nentwich and Kelan, 2014, p. 126). Thereby, masculinity can be performed – in private and public domains – by both men and women. So when masculine modes of behaviour and sociability are taken as the norm, women also aspire to, are assessed by, and need to align to the reproduction of such performative traits (Charles, 2014, p. 368; Shrestha, Joshi and Clément, 2019, p. 134ff).

Such performances of gender lead to inequality and constrain transitions because they are linked to activating and re-establishing symbolic hierarchies as well as structural arrangements. Whatever becomes gendered as masculine receives a higher status and is perceived as more professional as well as representing competence, while at the same time it reinforces the denial of the ethics of caring and distributive justice (Shrestha, Joshi and Clément, 2019, p. 148). For this reason, "doing gender" means "doing hierarchies" (Nentwich and Kelan, 2014, p. 127).

2.4 Conclusions

In this chapter, we have made an overdue start on engendering TM and the ST field. For this, a multi-facetted conceptualization of gender was introduced. Rather than reducing gender to an individual characteristic or even a synonym for women, it depicts structural, symbolic, and performative qualities that fundamentally affect the arrangement of societal structures, meaning systems, and interactions – and therefore the very subject matter and context of sustainability interventions by design, such as TM.

By discussing the influence that gender exerts on the different phases (setting the scene, framing challenges, formulating visions) and the questions and outcomes they imply, as well as the sites (team, arena, experiment) of the TM process, it has become clear that its theorizations of and interventions in STs are overlooking a vital imprint of targeted systems. This risks unintended outcomes as well as failure, while equally missing crucial opportunities in terms of ST. Understanding the whole gendered picture, however, can enable meaningful and effective measures leading to ecologically and socially sustainable outcomes.

ST researchers and practitioners engaged in related processes are therefore well advised to enquire about the gendered characteristics of the socio-technical system structures they address. Binary logics are embedded in existing as well as envisioned in-/formal institutions, organizations, or infrastructures, and contribute to the re-/configuration or de-/stabilization of entire systems. In particular, the gendered public/private-divide detaching masculinized production from feminized reproduction can prevent a comprehensive rethinking of economic processes and arrangements for a post-growth and post-carbon world.

In this, the role of gendered discourses in reinforcing, delegitimizing, or designing systemic structures should not be underestimated. It is therefore necessary to address linguistic (metaphors, narratives, etc.) as well as non-linguistic (images, icons, etc.) expressions and representations as part of socio-technical systems and transition processes with regard to masculinity/femininity because particular configurations and interventions acquire meaning and legitimacy through gendered symbolic orders. For instance, if climate change is simply framed as a techno-scientific issue, it seems only obvious to leave the solutions required to traditional masculine realms and logics.

Similarly, more analytical and design sensitivity is needed concerning gendered identities and their performative character. Because through them hierarchical structures and symbolic norms are re/produced on an interpersonal level, which can – in the end – impede the acceptance and realization of transition efforts. Among other things, the logics of "doing masculinity" can undermine feminized practices of care.

These steps towards an engendered TM can be further strengthened by adopting an intersectionality perspective, as discussed earlier. Such an approach is not limited to TM but can equally be applied in the wider context of ST studies, thereby enhancing the field in terms of its objectives to understand system transitions and guide them towards sustainable futures. At the same time, this approach also calls for more specific attention, guidance, and funding criteria in the context of research policy in order to ensure that beyond questions of equal representation, gender forms an integral part of the research approach and design. This appears to be indispensable if questions concerning societal transitions and sustainability are at stake.

References

Acker, J. (2004) 'Gender, capitalism and globalization', *Critical Sociology*, 30(1), pp. 17–41. doi:10.1163/156916304322981668

Arora-Jonsson, S. (2014) 'Forty years of gender research and environmental policy: Where do we stand?', *Women's Studies International Forum*, 47, pp. 295–308. doi:10.1016/j.wsif.2014.02.009

van den Bergh, J.C.J.M., Truffer, B. and Kallis, G. (2011) 'Environmental innovation and societal transitions: Introduction and overview', *Environmental Innovation and Societal Transitions* 1(1), 1–23. doi:10.1016/j.eist.2011.04.010

van den Bosch, S. and Rotmans, J. (2008) *Deepening, Broadening and Scaling Up. A Framework for Steering Transition Experiments*. Delft/Rotterdam: Knowledge Centre for Sustainable System Innovations and Transitions.

Boyd, S.B. (1997) 'Challenging the Public/Private Divide: An Overview', in Boyd, S.B. (ed.) *Challenging the Public/Private Divide*. Toronto: University of Toronto Press. doi:10.3138/9781442672819-003

Butler, J. (1990) *Gender Trouble: Feminism and the Subversion of Identity*. New York: Routledge (Thinking gender).

Butler, J. (2006) *Undoing Gender*. London: Routledge. Available at: www.dawsonera.com/depp/reader/protected/direct/SearchResults,buttons.eBookView.sdirect?sp=S9780203499627 (Accessed: 21 April 2021).

Charles, N. (2014) 'Doing gender, practising politics: Workplace cultures in local and devolved government', *Gender, Work & Organization*, 21(4), pp. 368–380. doi:10.1111/gwao.12042

Cielemęcka, O. and Daigle, C. (2019) 'Posthuman sustainability: An ethos for our anthropocenic future', *Theory, Culture & Society*, 36(7–8), pp. 67–87. doi:10.1177/0263276419873710

Crenshaw, K. (1991) 'Mapping the margins: Intersectionality, identity politics, and violence against women of color', *Stanford Law Review*, 43(6), p. 1241. doi:10.2307/1229039

Dashper, K. and Finkel, R. (2020) '"Doing gender" in critical event studies: A dual agenda for research', *International Journal of Event and Festival Management*, 12(1), pp. 70–84. doi:10.1108/IJEFM-03-2020-0014

Degele, N. (2008) *Gender, Queer Studies eine Einführung*. Stuttgart: UTB.

European Environment Agency (2019) *Sustainability transitions: policy and practice*, 9. Available at: https://doi.org/10.2800/641030 (Accessed: 23 May 2022).

Faulkner, W. (2000) 'The power and the pleasure? A research agenda for "making gender stick" to engineers', *Science, Technology, & Human Values*, 25(1), pp. 87–119. doi:10.1177/016224390002500104

Foster, E.A. (2017) 'Gender, Environmental Governmentality, and the Discourses of Sustainable Development', in MacGregor, S. (ed.) *Routledge Handbook of Gender and Environment*. 1st edn. Abingdon, Oxon; New York, NY: Routledge, 2017, pp. 216–228. doi:10.4324/9781315886572

Frantzeskaki, N., Hölscher, K., Bach, M., et al. (eds) (2018) *Co-creating Sustainable Urban Futures. A Primer on Applying Transition Management in Cities*. Cham: Springer International Publishing (Future city, 11). Available at: http://search.ebscohost.com/login.aspx?direct=true&scope=site&db=nlebk&db=nlabk&AN=1682241 (Accessed: 13 January 2018).

Frantzeskaki, N., Hölscher, K., Wittmayer, J.M., et al. (2018) 'Transition Management in and for Cities: Introducing a New Governance Approach to Address Urban Challenges', in Frantzeskaki, N. et al. (eds) *Cocreating Sustainable Urban Futures*. Cham: Springer International Publishing (Future City), pp. 1–40. doi:10.1007/978-3-319-69273-9_1

Harding, S.G. (1986) *The Science Question in Feminism*. Ithaca: Cornell University Press.

Haug, F. (2005) 'Gender relations', *Historical Materialism*, 13(2), pp. 279–302. doi:10.1163/1569206054127192

Hawkesworth, M. (2005) 'Engendering political science: An immodest proposal', *Politics & Gender*, 1(01), 141–156. doi:10.1017/S1743923X0523101X

Hendriks, C. (2009) 'Policy design without democracy? Making democratic sense of transition management', *Policy Sciences*, 40(2), pp. 341–368. doi:10.1007/s11077-009-9095-1

Hofmeister, S., Katz, C. and Mölders, T. (2012) 'Grundlegungen im Themenfeld Geschlechterverhältnisse und Nachhaltigkeit', in *Geschlechterverhältnisse und Nachhaltigkeit. Die Kategorie Geschlecht in den Nachhaltigkeitswissenschaften*. Opladen, Berlin, Toronto: Verlag Barbara Budrich, pp. 33–76.

Hölscher, K., Avelino, F. and Wittmayer, J. (2018) 'Empowering Actors in Transition Management in and for Cities', in Frantzeskaki, N. et al. (eds) *Cocreating Sustainable Urban Futures*. Cham: Springer International Publishing (Future City), pp. 131–158. doi:10.1007/978-3-319-69273-9_6

Hultman, M. (2017) 'Exploring Industrial, Ecomodern, and Ecological Masculinities', in MacGregor, S. (ed.) *Routledge Handbook of Gender and Environment*. 1st edn. Abingdon, Oxon; New York, NY: Routledge, 2017, pp. 239–252. doi:10.4324/9781315886572

Jhagroe, S. and Loorbach, D. (2015) 'See no evil, hear no evil: The democratic potential of transition management', *Environmental Innovation and Societal Transitions*, 15, pp. 65–83. doi:10.1016/j.eist.2014.07.001

Johnson, D. (2010) 'Sorting out the Question of Feminist Technology', in Layne, L.L., Vostral, S.L., and Boyer, K. (eds) *Feminist Technology*. Urbana: University of Illinois Press (Women, gender, and technology), pp. 36–54.

Kaijser, A. and Kronsell, A. (2014) 'Climate change through the lens of intersectionality', *Environmental Politics*, 23(3), pp. 417–433. doi:10.1080/09644016.2013.835203

Kenis, A., Bono, F. and Mathijs, E. (2016) 'Unravelling the (post-)political in transition management: Interrogating pathways towards sustainable change', *Journal of Environmental Policy & Planning*, 18(5), pp. 568–584. doi:10.1080/1523908X.2016.1141672

Kronsell, A. (2013) 'Gender and transition in climate governance', *Environmental Innovation and Societal Transitions*, 7, pp. 1–15. doi:10.1016/j.eist.2012.12.003

Kronsell, A. (2017) 'The Contribution of Feminist Perspectives to Climate Governance', in Buckingham, S. and Masson, V. le (eds) *Understanding Climate Change Through Gender Relations*. London; New York: Routledge, Taylor & Francis Group (Routledge studies in hazards, disaster risk, and climate change), pp. 104–120.

Lloro-Bidart, T. and Finewood, M.H. (2018) 'Intersectional feminism for the environmental studies and sciences: Looking inward and outward', *Journal of Environmental Studies and Sciences*, 8(2), pp. 142–151. doi:10.1007/s13412-018-0468-7

Loorbach, D. (2010) 'Transition management for sustainable development: A prescriptive, complexity-based governance framework', *Governance: An International Journal of Policy, Administration, and Institutions*, 23(1), pp. 161–183.

Loorbach, D., Frantzeskaki, N. and Avelino, F. (2017) 'Sustainability transitions research: Transforming science and practice for societal change', *Annual Review of Environment and Resources*, 42(1), pp. 599–626. doi:10.1146/annurev-environ-102014-021340

Loorbach, D. and Rotmans, J. (2010) 'The practice of transition management: Examples and lessons from four distinct cases', *Futures*, 42(3), pp. 237–246. doi:10.1016/j.futures.2009.11.009

MacGregor, S. (2009) 'A stranger silence still: The need for feminist social research on climate change', *The Sociological Review*, 57(2_suppl), pp. 124–140. doi:10.1111/j.1467-954X.2010.01889.x

MacGregor, S. (2010) 'Plus ça (Climate) Change, Plus c'est la même (Masculinist) Chose: Gender Politics and the Discourses of Climate Change', in Lee, Y.-L. (ed.) *The Politics of Gender a Survey*. London: Routledge, pp. 83–101.

McCall, L. (1992) 'Does gender fit? Bourdieu, feminism, and conceptions of social order', *Theory and Society*, 21(6), pp. 837–867. doi:10.1007/BF00992814

Meadowcroft, J. (2009) 'What about the politics? Sustainable development, transition management, and long term energy transitions', *Policy Sciences*, 42(4), pp. 323–340. doi:10.1007/s11077-009-9097-z

Nentwich, J.C. and Kelan, E.K. (2014) 'Towards a topology of "doing gender": An analysis of empirical research and its challenges', *Gender, Work & Organization*, 21(2), pp. 121–134. doi:10.1111/gwao.12025

Ourahmoune, N., Binninger, A.-S. and Robert, I. (2014) 'Brand narratives, sustainability, and gender: A socio-semiotic approach', *Journal of Macromarketing*, 34(3), pp. 313–331. doi:10.1177/0276146714528335

Paulus, S. (2015) 'Methodologische Überlegungen und methodisches Vorgehen bei einer intersektionalen Dispositivanalyse', *Forum Qualitative Sozialforschung*, 16(1), 1–26. Available at: http://nbn-resolving.de/urn:nbn:de:0114-fqs1501210.

Prügl, E. (2021) 'Untenable dichotomies: De-gendering political economy', *Review of International Political Economy*, 28(2), pp. 295–306. doi:10.1080/09692290.2020.1830834

Roorda, C. et al. (2014) *Transition Management in the Urban Context. Guidance Manual*. Rotterdam: DRIFT, Erasmus University Rotterdam, p. 49.

Rotmans, J. and Loorbach, D. (2009) 'Complexity and transition management', *Journal of Industrial Ecology*, 13(2), pp. 184–196. doi:10.1111/j.1530-9290.2009.00116.x

Sauer, B. (2018) 'Intersectionality', *Krisis. Journal for Contemporary Philosophy*, 2, pp. 86–87.

Scantlebury, K. (2014) 'Gender Matters: Building on the Past, Recognizing the Present, and Looking Toward the Future', in Abell, S.K. and Lederman, N.G. (eds) *Handbook of Research on Science Education*. New York; London: Routledge, Taylor & Francis Group, pp. 187–203.

Shove, E. and Walker, G.P. (2007) 'Caution! Transitions ahead: Politics, practice and transition management.', *Environment and Planning A*, 39(4), pp. 763–770

Shrestha, G., Joshi, D. and Clément, F. (2019) 'Masculinities and hydropower in India: A feminist political ecology perspective', *International Journal of the Commons*, 13(1), p. 130. doi:10.18352/ijc.920

Standal, K., Talevi, M. and Westskog, H. (2020) 'Engaging men and women in energy production in Norway and the United Kingdom: The significance of social practices and gender relations', *Energy Research & Social Science*, 60, p. 101338. doi:10.1016/j.erss.2019.101338

Walby, S. (2020) 'Varieties of gender regimes', *Social Politics*, 27(3), pp. 414–431. doi:10.1093/sp/jxaa018

Wallach Scott, J. (2010) 'Gender: Still a useful category of analysis?', *Diogenes*, 57(1), pp. 7–14. doi:10.1177/0392192110369316

West, C. and Zimmerman, D. (1987) 'Doing gender', *Gender & Society*, 1(2), pp. 125–151. doi:10.1177/0891243287001002002

Winker, G. and Degele, N. (2011) 'Intersectionality as multi-level analysis: Dealing with social inequality', *European Journal of Women's Studies*, 18(1), pp. 51–66. doi:10.1177/1350506810386084

Wittmayer, J. et al. (2018) 'Transition Management: Guiding Principles and Applications', in Frantzeskaki, N. et al. (eds) *Co-creating Sustainable Urban Futures*. Cham: Springer International Publishing (Future city, 11), pp. 81–101. Available at: http://search.ebscohost.com/login.aspx?direct=true&scope=site&db=nlebk&db=nlabk&AN=1682241 (Accessed: 13 January 2018).

Wolfram, M. and Kienesberger, M. (2021) 'Gender in sustainability transition studies: Concepts, blind spots and future orientations', *Environmental Innovation and Societal Transitions* [Preprint].

Youngs, G. (2000) 'Breaking Patriarchal Bonds. Demythologizing the Public/Private', in Marchand, M.H. and Runyan, A.S. (eds) *Gender and Global Restructuring: Sightings, Sites and Resistances*. London; New York: Routledge (RIPE series in global political economy), pp. 44–58.

3

GENDER WITHIN SUSTAINABLE DECISIONS

A 3-D sustainability perspective

Volker Mauerhofer

3.1 Introduction

The United Nations have recently repeatedly re-emphasized the three dimensions of sustainable development (SD), the environmental, the social, and the economic ones (UN 2012, 2015). Gender equality is mainly reflected in the social dimension; that has been during the past the one least addressed by academia (Lehtonen, 2004; Mauerhofer, 2013). Nevertheless, since the early 90s, gender gained an increasing attention among scholars. Allen and Sachs (1991) emphasized the social side of SD as class, gender and race indicating the wider scope of this dimension. Another author outlined in the same year some of the gender implications for SD in order to prepare for the UNCED Earth Summit in Brazil 1992 (Dankelman, 1991). Jacobson (1992) points out the gender bias existing in the Global South in food production and management of resources as a roadblock to SD and thereby highlights the connections of gender with the environmental and economic dimension of SD. The United Nations announced 2012 Sustainable Development Goals (SDGs) and introduced them in 2015 consecutively to the Millennium Development Goals (UN 2012, 2015). "Gender" has been mentioned in the resolution "The Future We Want" from 2012 for 26 times in total throughout the document (UN 2012) and then focused upon within the resolution from 2015 in Goal 4 dealing with education and learning and in the horizontal gender Goal 5 (UN 2015). Within Goal 5, it is stated that: "Gender equality is not only a fundamental human right, but a necessary foundation for a peaceful, prosperous and sustainable world." While gender is obviously an issue most strongly related to the social dimension of an SD, its relation and inter-connectedness to the other two dimensions are often overlooked and less assessed due to its complexity.

DOI: 10.4324/9781003174257-4

This chapter addresses the question in how far this relation and interconnectedness of gender can be addressed by means of the normative criteria of 3-D Sustainability (Mauerhofer 2008, 2019).

This chapter starts in the following with a short introduction into 3-D Sustainability as conceptual assessment framework and describes in detail especially its six normative criteria further part of the analytical framework applied. The findings provide examples of gender and sustainability studies found during the in-depth literature review for each of the six normative criteria of 3-D Sustainability. The allocation of these examples to the criteria are then discussed respectively and the sustainability implications outlined. A summary of main findings and the related conclusions finalize the chapter.

3.2 Theoretical framework and methodology

This chapter fosters the idea of an SD consisting of balancing and integrating multifaceted interests within and among its environmental, social, and economic dimensions already included in the famous "Brundtland Report" (WCED 1987) and reconfirmed recently in the Rio+20 outcome documents (UN 2012) and its following SDGs (UN 2015) such as those already described above. The chapter applies as theoretical framework a concept called "3-D Sustainability" (Mauerhofer, 2008, 2019). This concept has been developed to provide orientation for decision-making in complex sustainability contexts and therefor seems to be appropriate to assess the relation and interconnectedness of gender within all the three dimensions of an SD (Figure 3.1).

3-D Sustainability differentiates first in more descriptive way among environmental, social, and economic capital and (carrying) capacity and then provides for the interconnections among them six more normative criteria within a preliminary fixed hierarchy (Figure 3.1). The six normative criteria can be applied for flexible balancing and integrating the three dimensions of an SD and are defined in absolute and relative manners as well as characterized by particular features of their objectives (Table 3.1).

The preliminary hierarchy of the criteria for decision-making expresses by its flexibility also the closeness of the six criteria. The six criteria also refer with differing intensities to the ecologically sustainable scale; the socially fair distribution; and the economically efficient allocation of resources, such as defined by Costanza et al. (1997: 80, 83) and Daly (1992). According to these authors, these three features can be addressed in this exact order by decision-makers by using systems of property rights and transfers among different stakeholders for distribution and through market-based mechanisms to efficiently allocate resources among alternative products (see also Mauerhofer, 2012: 653f) (Figure 3.2).

Gender equality is mainly reflected in the social dimension such as already mentioned in the Introduction section, and therefore related to a socially fair distribution within Figure 3.2 among sexes (but not classes and races). Despite, Figure 3.2 shows for each of three big arrow also side-arrows. These indicate the connections

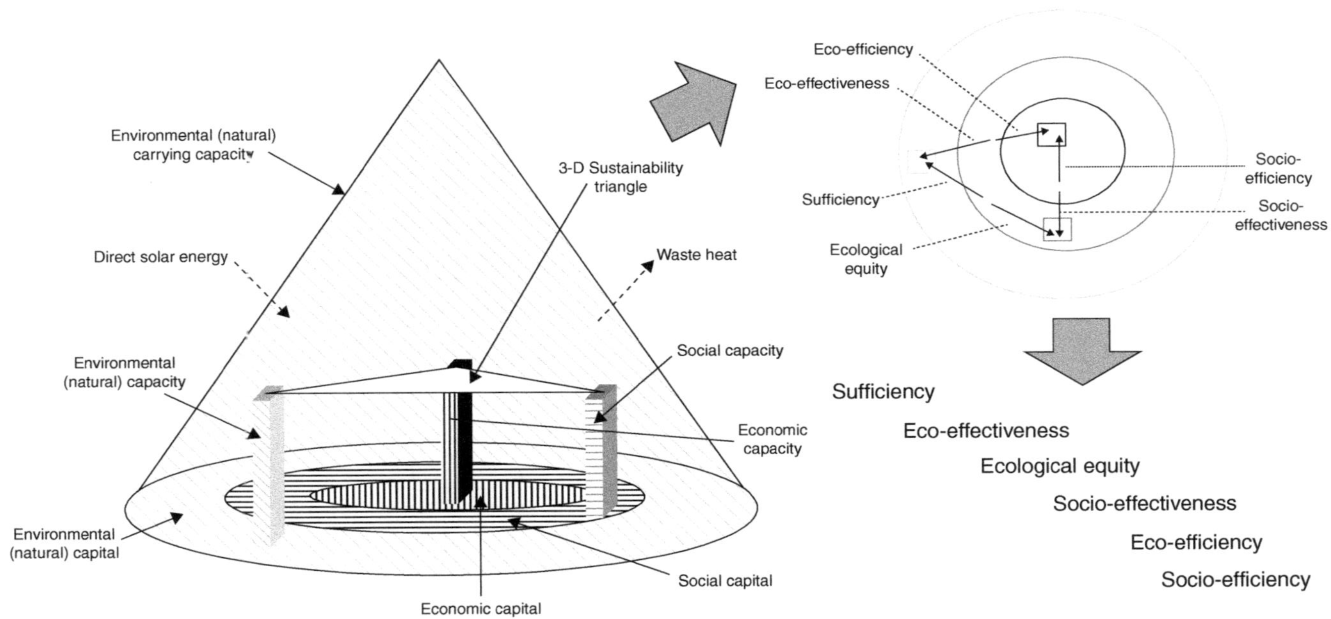

FIGURE 3.1 3-D Sustainability in detail and from above (adapted from Mauerhofer 2008a; 2019a,b; triangle based on Dyllick and Hockerts, 2002).

TABLE 3.1 Definitions and features of the six normative criteria of 3-D sustainability

Criteria	*Definitions*	*Characteristic features of objectives*
Sufficiency	Individual's voluntary decision to consume/produce no unit (more) or even unit's less	Absolute, quantitative and/or qualitative, but sustainability depends on current consumption/production level;
Eco-effectiveness	Capacity of an action to achieve its expressed ecological objectives	Absolute, at least >50% improvement, no rebound effect
Ecological equity	Intra- and intergenerationally acting under the "veil of ignorance" (Rawls 1971)	
Socio-effectiveness	Capacity of an action to achieve its expressed societal objectives	
Eco-efficiency	Input and output ratio (per time) of energy/waste for product/service	Relative, 0–99% improvement per unit (product/service/person), rebound effect in longer run
Socio-efficiency	Input and output ratio (per time) of an action concerning persons	

Adapted from Mauerhofer 2019.

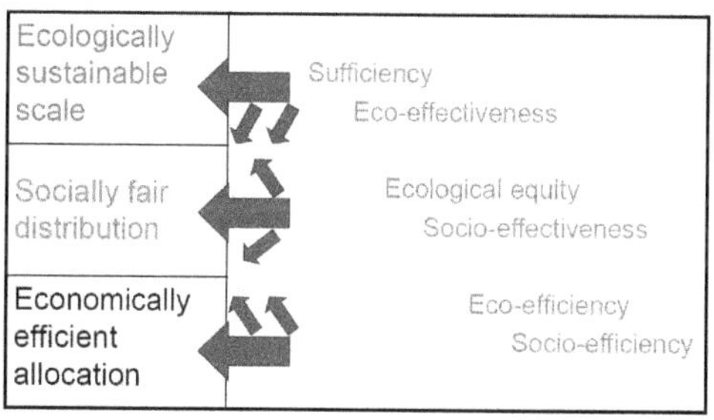

FIGURE 3.2 Rough external relationship between the six criteria of 3-D sustainability and scale, distribution and allocation (adapted from Mauerhofer 2012; 2019).

to the two other overall goals respectively and also to the other criteria of 3-D Sustainability. These interrelations are also discussed in the following chapters for each of the six criteria of 3-D Sustainability as a result from the in-depth literature review. While publications which look on gender from a purely descriptive view on capitals or capacities are not further elaborated (see, e.g. Manjengwa, 1998; Franz-Balsen, 2010).

The literature review has been implemented by the end of 2019 by means of the most advanced scientific databases Web of Science (the option "All Databases" was selected therein) and Scopus, respectively. Title search has been applied for the terms "gender" and "sustainab*," and the literature gained in this way has been screened according to in how far this relation and interconnectedness of gender can be addressed by means of the normative criteria of 3-D Sustainability.

3.3 Findings

In the following, six subchapters for the six normative criteria of 3-D Sustainability are presented, and the related literature is respectively summarized and discussed.

While each of the selected pieces of literature naturally has a strong relation to ecological equity in the sense of inter- and intrageneration equity, only those gender and sustainability related examples found were allocated to that criterion which did not show any stronger relation to one of the other five criteria.

3.3.1 Gender and sufficiency

Ahonsi (1995) already early suggests for Africa that a transition to lower fertility and higher women's status may be this continent's most realistic road to SD. The earlier suggestion can be seen closely connected to sufficiency if it is considered as a voluntary act and dependent on the starting level for that reduction. Similarly, Tisdell (2001) forecasts that a reduction of population growth – through provision of sufficient human capital for females as a crucial aspect – is likely to have favourable consequences for SD.

3.3.2 Gender and eco-effectiveness

Thomas-Slayter and Rocheleau (1994) already are linking gender to effective natural resource management and SD by arguing that attention to gender is central to increasing the equity and effectiveness of local-level management of natural resources.

In this sense, Smetschka et al. (2009) assess structural change in agriculture with its ecological, economic, and social aspects by means of an agent-based model with the farm households of two Austrian villages as agents. Therein, it was possible to stop the problem of decreasing farm activities and increasing forests in the sustainability scenario while the workload on women farmers concomitantly increased. These results indicate a trade-off wherein it is questionable whether it is socially effective or at least efficient in sustainability terms.

Cuellar-Gomez (2009) analysed how gender-based marketing strategies for Colombian coffee have affected gender roles and the economic and social advancement of women producers. The author concludes that the use of gender in global marketing can contribute to the economic, social as well as environmental sustainability of the communities and persons involved. This result indicates a potential for eco-effectiveness as well as socio-effectiveness.

While Meinzen et al. (2014) found with regard to gender, sustainability, and conservation that – despite early claims according to which women are naturally more conserving of resources – the empirical literature especially gives a more mixed and nuanced picture of multiple influences on conservation. Similarly, McClanahan and Abunge (2017) concluded from their empirical analysis of traders in 19 Kenyan coral reef fisheries landing sites distributed among four gear management systems that gender coexistence is most likely to be achieved by managing for intermediate resource levels where net production, catch, and fish body size diversity are high and that the reduction of risk and increasing the capital and mobility of women traders should reduce their chances of exclusion when fisheries are sustainable.

Other research concentrates more on climate effective approaches. Thus, Broeckhoven (2015) explored linkages between gender equality, women's rights, and climate change by exploring its international legal and policy dimension and concludes that attempts to integrate both perspectives have taken a parallel rather than an integrated approach, highlighting finally the need for a gender-responsive human rights-based approach to climate change. Miralles-Guasch et al. (2016) empirically found from a large travel survey taken in 2006 in Spain that women are using sustainable transport modes more often than men, and travelling for more diverse reasons indicating gender is thus a fundamental variable in terms of energy consumption and the emission of greenhouse gases.

In summary, these results show several example where increase in gender considerations have lead to more eco-effective results while also some examples exist that are indicating the opposite and stating reasons therefore.

3.3.3 Gender and ecological equity

3-D Sustainability distinguishes within ecological equity between intragenerational and intergenerational equity. Already in 1997, Zein-Elabdin (1997) places gender within the SD paradigm as a matter of intra-generational equity requisite for sustainability, when criticizing the World Bank for its tendency to disproportionately implicate women in deforestation.

Gender equality as a sustainability-related research object in order to promote the involvement of and results achieved by females has since then in general entered a wide range of scientific thematic assessments such as in the area of Corporate Social Responsibility (e.g. Grosser, 2009), mobility and transport (e.g. Hanson, 2010, Turner, 2012), religion (e.g. Rozario and Samuel, 2010), unpaid work (Fudge, 2011, Coote, 2012), energy (Oparaocha and Dutta, 2011), and risk perceptions (Vu, 2012).

Findings from the literature assessed in term of status and perspectives for intragenerational equity show a mixed picture.

In terms of status, Adaramaja et al. (2010) empirically investigated gender as a determinant of individual lifestyle for SD in a state of Nigeria. Their results indicate that gender does not influence or determine respondents' smoking habit, alcoholic lifestyle, nutritional habit, and unsafe sexual practices but that significant differences exist in male and female drug habit and physical inactivities (sedentary lifestyle).

In terms of perspectives, Lisboa and Lusa (2010) concluded from a brief comparative analysis of rural women's role in food production and natural resource management in Brazil, Mexico, and Cuba that the development model in all three countries (two capitalist and one communist ones) still prioritizes the male figure in relation to land tenure, access to credit, and purchase of equipment or other material resources. When it comes to a positive youth development perspective, a cross-sectional designed study with 858 Ghanaian adolescents attending senior high school (44% girls) resulted that boys scored higher on competence relative to girls and that girls scored higher on caring relative to boys while no significant differences were found for socio-economic background. (Wiium et al., 2019). The authors assume that cultural norms and socialization processes perpetuate gender inequality and may be important to understanding the observed gender differences.

In terms of inter-generational sustainability, Mwita and Murphy (2017) point out the complex drivers of educational exclusion and non-progression of female adolescents in rural Tanzania as – despite targeted government policies, donor funding, and multiple development interventions – this area has faced declining rates of academic progression for young girls over the past decade.

Olsson and Gericke (2017) empirically identified in a sample of 2,413 Swedish Students aged 12–19 a gender gap in students' sustainability consciousness that increases throughout the age span and is amplified in schools oriented towards Education for Sustainable Development.

Najjar et al. (2013) provide an insightful case study related to learning about sustainability and gender from Kenya. Through Farmer Field Schools in the Taita Hills there, a significant change in meaning perspective has occurred among men who had overcome personal biases and a cultural practice of land inheritance for males which then have included also their daughters.

Dür and Keller (2018) showed in a case study the effect of education for SD through a qualitative content analysis of Austrian and Indian students' texts regarding the SDGs "Gender Equality" and "Decent Work and Economic Growth" because the students' awareness increased, and their evaluation of topics related to the discussed SDGs changed due to their international collaboration.

Another intergenerational example is provided by Remington-Doucette and Musgrove (2015). When presenting the results of a university classroom assessment that aimed at determining the extent to which five key sustainability competencies develop in students during an introductory transdisciplinary sustainability course, they found inter alia that females improved interpersonal competence more than males.

These intergenerational examples all indicate into the same direction, namely an increase of ecological equity (in the sense of 3-D Sustainability) when it comes to education and gender for the next generation.

3.3.4 Gender and socio-effectiveness

De-Luis et al. (2010) explore the current vision whereby Spanish companies have formed their Corporate Social Responsibility policy in terms of particular social (board's gender balance) and environmental stakeholder issues. While currently the financial approach was found to be the dominant one, existing Spanish law might be modulating this approach and facilitating a normative approach in order to help companies to achieve accountability to society as regards sustainability. These results of De-Luis et al. (2010) indicate the necessity of a binding intervention in the sense of socio-effectiveness in order to implement and secure social (gender) sustainability goals.

In comparison, Galbreath (2011) investigated whether there is a link between women on boards of directors and corporate sustainability. Using a sample of publicly listed Australian firms, this study suggests some level of support that a link does exist while it also found that women directors are not significantly associated with environmental quality.

In terms of the status of socio-effectiveness, Huang and Ali (2015) found in a rapid evaluation about the impact of leading industries of mining and tourism on sustainability that mining areas performed better than tourism areas in economic aspects but fell behind in social development, especially regarding the issue of gender balance. The results of Brahmapurkar (2017) show a more negative picture. This author indicates in a study based on data from National Family Health Survey 2015–16 in India from15 states and 3 Union Territories that in a rural area, all states except Meghalaya and Sikkim had the significantly higher percentage of illiterate women as compared to men and that Bihar and Madhya Pradesh had higher illiterate women (53.7% and 48.6%), as compared to men (24.7% and 21.5%), respectively ($P < 0.000$). A substantial disparity in terms of socio-effectiveness status reveals also the following study found. Gligorijević et al. (2016) provide results on gender differences for job sustainability related to self-employed home-based business (HBB), whereas in their sample of 310 HBB owners, 48.9% of women and 35.8% of men started their own business out of necessity.

In terms of perspective and trend, an example of socio-effectiveness in connection with gender has been recently presented by Arcanjo (2019). This author empirically analysed retirement pension reforms in six European social insurance schemes between 2000 and 2017 and demonstrates how the reforms have focused on strengthening the financial sustainability of systems but may also have contributed to even greater gender inequality in old age protection. If this assumption proves true, then a "negative socio-effectiveness" can be suggested in the frame of 3-D Sustainability. A similar negative trend was pointed out by Adegbite and Machethe (2020) for Nigeria's financial inclusion gender gap (FIGG) in smallholder agriculture. Their findings establish that the FIGG in smallholder agriculture was 12% in 2016, while considering the whole population; it increased from 7% in 2011 to 20% in 2014 and 24% in 2017.

The allocation to socio-effectiveness is not always that clear-cut because several studies overall relate its analysis not only to the social and economic dimension of sustainability but also to the environmental one.

But while socio-effectiveness according to the features allocated to it in 3-D Sustainability requires for a sustainable solution and "Absolute, at least >50% improvement, no rebound effect" (Table 3.1), several of the examples indicate the opposite. The "social scissor" is therefore opening – also in terms of gender – wider towards not sustainable perspectives, opposite to what 3-D Sustainability requires in terms of decision-making.

3.3.5 *Gender and eco-efficiency*

Eco-efficiency is strongly connected to economically efficient allocation of resources and energy within 3-D Sustainability and in this way closely connected to business and corporate sustainability. Weaven et al. (2013) already found in their empirical assessment in an undergraduate business school about specific perceptions of business sustainability issues, sustainability education, and the importance that students attach to sustainability in their choice of future employer, numerous significant differences between the female and male cohorts and also observed gender differences across the constructs predicting employment choice.

Also in connection with business schools, Miotto et al. (2019) observed in their empirical assessment of gender equality and UN SDGs related to priorities and correlations in the top business schools' communication and legitimation strategies that gender equality-related topics are a source of positive impact and legitimacy for top business schools. Both studies seem to focus on individual business, therefore prone to economic growth and also to rebound effects (on this, see e.g. Binswanger, 2001, Saey-Volckrick, 2020).

Coming from business education to practice, Slepian and Jones (2013) found in their empirical study that corporate women see sustainability-related concerns and values to be significantly more important to them personally than do their male colleagues and that they consider their companies' sustainability-related value activities, initiatives, and priorities from these foundational value and ethical orientations. In a similar direction, Kassinis et al. (2016) empirically assessed a sample of 296 US publicly traded firms over a five-year period. They found that both, firms with more gender diverse boards of directors ("demographic" gender diversity) and firms with more policies and practices that enable or reinforce gender diversity throughout the organization ("structural" gender diversity), significantly adopted more environmentally responsible policies and practices repectively. In this line, Al-Shaer and Zaman (2016) found in a study related to board gender diversity and sustainability reporting quality that gender diverse boards were associated with higher quality sustainability reports and independent female directors had greater effect on sustainability reporting quality than female directors. Also, Ben-Amar et al. (2017) found evidence – based on a sample of publicly listed Canadian firms over the period 2008–2014 – that the likelihood of voluntary climate change disclosure increases with women percentage on boards. Nadeem et al. (2017) similarly provide empirical evidence supporting a board diversity case for corporate sustainability performance based on a sample of Australian Securities Exchange listed firms (similar with an even broader sample from different continents Shoham et al., 2017). Valls Martínez et al. (2019) did similar empirical tests using the data from top Spanish listed companies for the period 2003 to 2017. Their findings indicate that the female presence in management positions is positively linked to a voluntary disclosure of Corporate Social Responsibility reports and the inclusion in a sustainability index (similar for Malaysia, Zahid et al., 2020).

While Mapparessa et al. (2017) found in their empirical analysis of factors effecting sustainability report disclosure of 22 companies listed in Indonesia Sustainability Reporting Award that political visibility which is measured by company size has a negative and significant effect to sustainability report disclosure, while characteristics measured by type of company and gender diversity have no effect.

Regarding the audit of reports a mixed picture can be found. Bravo and Reguera-Alvarado (2019) assess the connection between female representation on audit committees (ACs) and specific information attributes of environmental, social, and governance (ESG) disclosures. Their results show a positive association between gender diversity in the AC and the quality of voluntary ESG reporting, which results in greater comprehensiveness and relevance. They further indicate that the busyness of the AC negatively moderates the influence of female AC members.

Furlotti et al. (2019) also gained mixed results in their empirical assessment about whether the presence of women with different responsibilities is associated with gender policies disclosed in corporate social responsibility or sustainability reports by using the data of listed companies in Italy. While they found a positive association between the presence of women in the role of chairperson and the implementation with disclosure of gender policies, different results were gained when the women were in the position of chief executive officer, indicating a non-significant perception of female problems when a woman occupies that position.

Augustine et al. (2016) found related to gender diversity within the workforce in the microfinance industry in Africa in an analysis of a data set of 1,389 observations in 1,053 firms suggests that gender diversity enhances economic performance, especially in Africa.

Meek and Sullivan (2018) explored potential impact factors on environmental sustainability orientation with a sample of 208 existing entrepreneurs. Their findings suggest that no gender differences exist but that organizational identity and organizational tenure do have a significant impact.

Leaving cooperate topics now behind, Soares (2006) explained for a municipality in the Altos de Chiapas (Mexico) the factors (socio-cultural, economical, climatic, and technical) that promote or hinder the acceptance of alternative technologies that aim to reduce the consumption of firewood among rural families and thus decrease the female workload and increase the conservation of the woods. However, the issue of a potential population increase and increase of the numbers of families – which could lead to a rebound effect in environmental terms – was apparently not addressed.

Previous global gender-related policy approaches do not always seem to be effective on the regional level.

Thus, Vyas-Doorgapersad (2014) revealed regarding the relationship among Millenium Development Goal – MDG 3 (about gender equality and women empowerment) and MDG 8 (about ICT information and communication technologies) that there is a lack of gender-responsive ICT in West African countries and provide policy recommendations to improve the situation. O'Manique and Fourie (2016) generally criticize the SDGs to perpetuate rather than challenge the systemic drivers of gender injustice while liberal feminism and its more insidious twin, economic neoliberalism, have captured mainstream development discourse according to their opinion. Koehler (2016) analyses whether the SDGs contain policies for gender equity and equality as well as for SD in tune with planetary boundaries but found that they are not ambitious and lack a systematic policy approach, while proposing that they can be used creatively and subversively – by incorporating other, more progressive conventions and declarations – to move towards gender and climate justice. Similarly, Goetz and Jenkins (2016: 127) enumerate some reasons why in the SDG 16 "peace and governance" none of the targets included specific measures to ensure women's equal participation in governance institutions and peace processes. Among them are "the relatively strong governance orientation of the gender-equality goal (Goal 5), the political tensions surrounding Goal 16 prior to its adoption, the compression necessitated by the merging of what had originally been two separate goals (on peace and governance, respectively), and the 2030 Agenda's tendency to focus on ends rather than means." Agarwal (2018) raises similar concerns on SDG 5 in its relation to other SDGs (1, 2, 13–15). However, these authors do not elaborate about the practical consequences for the SDG text length if the essence of every SDG would be mentioned in each other SDG. In comparison, Manandhar et al. (2018) propose concerted and collaborative actions on gender, health and the 2030 Agenda for SD to connect gender (SDG 5) and health (SDG 3) with 13 other SDGs.

Finally coming to eco-efficient product standards, Chiputwa and Qaim (2016) explored inter alia the relationship between sustainability standards. They found that that certification increased calorie and micronutrient consumption, mainly through higher incomes and improved gender equity, as well as that in certified households, women had greater control of coffee production and monetary revenues from sales.

In summary, a wide range of literature exists related to gender and eco-efficiency. It is the topic in the heart of businesses that relatively address single units of production and consumption but usually is not discussed in connection to limits to production consumption numbers, particularly in the Global North, and therefore easily prone to rebound effects.

A wide array of literature also exists in terms of gender and SDGs where the gender-related SDG 5 is widely criticized, and additional approaches are advocated.

3.3.6 Gender and socio-efficiency

Literature in connection with gender and sustainability allocated in the following to socio-efficiency is often also rather hard to be allocated to that criteria as often environmental dimension is somehow also involved.

Palmer Jones and Jackson (1997) examine the question of how far small-scale, intermediate technology based on energy-intensive human work is compatible with SD objectives emphasizing gender equity. Based on a case study about treadle pumps for irrigation in Bangladesh, they illustrate trade-offs between development strategies based on "labour intensive growth" and the well-being of men and women and argue for more rigorous analyses of gender divisions of labour, which include work intensity in combination with time allocation. Similarly, Wickramasinghe (2003) reported in a field study in Sri Lanka related to biomass supply that health risks and the self-reported ailments are disproportionately distributed towards women and the repercussions suffered by women in their effort of ensuring household energy security are serious.

Similarly health-related, Buckingham-Hatfield (2002) points out a need to improve the worse (relatively compared to men) women's health, education, economic position, and involvement in environmental decision-making and concludes after assessing for progress that, while some improvements appear to have been made, these are by no means uniform across time and space as well as that persistent inequalities continue to exist.

Galmiche-Tejeda and Townsend (2006) implemented a case study on the extension for Semi-Subsistence Fish Farming in Tabasco (Mexico) related to SD and gender hierarchies. They highlight that an appreciation of the different understandings of fish farming by different family members and its gendered micro-geography are the key issues in explaining the relative successes and failures in recent extension efforts, where historically governmental and non-governmental organizations have worked with individuals and groups of (usually men) farmers rather than family groups.

More related to the non-profit sector, Parida and Sinha (2010) assess the performance and sustainability of bank linkage programmes for self-help groups (SHGs) in India from a gender perspective. Overall, their performance analysis reveals – based on data from a survey covering six states – that all-female SHGs perform best, particular in terms of recovery of loans and per capita saving. This assessment was solely done on a group basis, and any limits of growth apparently have not been considered. Thus, future rebound effects from the funded activities cannot be excluded, at least in the long term.

Literature related to gender and sustainability that was allocated to socio-efficiency concern several times the issue of health but also multiple other topics with less connection to the environmental dimension.

3.4 Further discussion and Conclusions

The current study clearly shows that each of the six normative decision-making criteria can be connected to gender- and sustainability-related literature examples.

Least literature is found regarding the criterion that is placed highest in the flexible decision-making hierarchy of 3-D Sustainability, namely sufficiency. While apparently, the criteria that more literature could be allocated to constitute the economic-oriented ones.

3-D Sustainability is working with descriptive and normative features. The latter have been applied in this study to allocate literature preliminary identified to.

Of course, other approaches exist that in an overall way trying to systematize gender into sustainability.

In comparison, Graef et al. (2018) follow an overall impact-oriented approach when systemizing gender integration with rural stakeholders' sustainability impact assessments. Their results show how important it is to integrate female–male segregated assessments and perceptions before and while implementing food-securing upgrading strategies. Also, the World Health Organization developed and implemented various tools and approaches to operationalize the commitment to equity, gender, and human rights, particular oriented towards leaving no one behind in the SDGs (Zamora et al., 2018). While the question remains leaving no one behind *what*, especially but not only regarding the economic dimension of sustainability embedded in the environmental carrying capacity with 3-D Sustainability.

Asongu and Odhiambo (2020) empirically investigated how enhancing gender inclusion affects inequality in 42 African countries for the period 2004–2014 by using three inequality indicators, namely, the Gini coefficient, Atkinson index, and Palma ratio. The following main findings are established. They found positive net effects on inequality from the enhancement of gender inclusion dynamics based on two measurements, namely female labour force participation and female employment.

Such concrete indicators and measurements also need to be applied for further assessments of the absolute and relative normative and descriptive criteria within 3-D Sustainability.

In summary, the approach chosen can be considered innovative as the connection between 3-D Sustainability and gender has never been assessed before, to the best knowledge of the author. Due to its theoretical replicability, the approach has potential for a global application in terms of sustaining the social carrying capacity within 3-D Sustainability and could be in such a context further elaborated in a more systematic literature review and with a more finetuned allocation methodology.

References

Adaramaja, S.R., Adenubi, O.S., Nnbueze, U., 2010. Gender as a determinant of individual lifestyle for sustainable development in Africa. *Research Journal of Applied Sciences* 5, 27–31.

Adegbite, O.O., Machethe, C.L., 2020. Bridging the financial inclusion gender gap in smallholder agriculture in Nigeria: An untapped potential for sustainable development. World Development 127.

Agarwal, B., 2018. Gender equality, food security and the sustainable development goals. *Current Opinion in Environmental Sustainability* 34, 26–32.

Ahonsi, B.A., 1995. Gender relations, demographic change and the prospects for sustainable development in Africa. Africa development. *Afrique et développement* 20, 85–114.

Allen, P.L., Sachs, C.E., 1991. The social side of sustainability: Class, gender and race. *Science as Culture* 2, 569–590.

Al-Shaer, H., Zaman, M., 2016. Board gender diversity and sustainability reporting quality. *Journal of Contemporary Accounting & Economics* 12, 210–222.

Arcanjo, M., 2019. Retirement pension reforms in six european social insurance schemes between 2000 and 2017: More financial sustainability and more gender inequality? *Social Policy and Society* 18, 501–515.

Asongu, S.A., Odhiambo, N.M., 2020. How enhancing gender inclusion affects inequality: Thresholds of complementary policies for sustainable development. Sustainable Development 28, 132–142.

Augustine, D., Wheat, C.O., Jones, K.S., Baraldi, M., Malgwi, C.A., 2016. Gender diversity within the workforce in the microfinance industry in Africa: Economic performance and sustainability. *Canadian Journal of Administrative Sciences* 33, 227–241.

Ben-Amar, W., Chang, M., McIlkenny, P., 2017. Board gender diversity and corporate response to sustainability initiatives: Evidence from the carbon disclosure project. *Journal of Business Ethics* 142, 369–383.

Binswanger, M., 2001. Technological progress and sustainable development: What aboutthe rebound effect? *Ecological Economics* 36, 119–132.

Brahmapurkar, K.P., 2017. Gender equality in India hit by illiteracy, child marriages and violence: A hurdle for sustainable development. *Pan African Medical Journal* 28, 1–9. DOI: 10.11604/pamj.2017.28.178.13993

Bravo, F., Reguera-Alvarado, N., 2019. Sustainable development disclosure: Environmental, social, and governance reporting and gender diversity in the audit committee. *Business Strategy and the Environment* 28, 418–429.

Broeckhoven, N., 2015. Gender Equality, Women's Rights, and Environmental Sustainability The Case of Climate Change. In: Brems, E., Van Der Beken, C., Yimer, S.A. (Eds.), *Human Rights and Development: Legal Perspectives from and for Ethiopia*. Martinus Nijhoff; Lam edition pp. 360–382.

Buckingham-Hatfield, S., 2002. Gender equality: A prerequisite for sustainable development. *Geography* 87, 227–233.

Chiputwa, B., Qaim, M., 2016. Sustainability Standards, Gender, and Nutrition among Smallholder Farmers in Uganda. *Journal of Development Studies* 52, 1241–1257

Coote, A., 2012. Growing the core economy: Gender, time and sustainable development. *Local Economy* 27, 788–795.

Costanza, R., Daly, H.E., Goodland, R., Norgaard, R.B., 1997. *An Introduction to Ecological Economics*. St. Lucie Press, Boca Raton, Florida (275 pp).

Cuellar-Gomez, O.L., 2009. Gender and agricultural sustainability: Case study of Colombian coffee. *International Journal of Innovation and Sustainable Development* 4, 216–225.

Daly, H.E., 1992. Allocation, distribution and scale: Towards an economics that is efficient, just and sustainable. *Ecological Economics* 6, 185–193.

Dankelman, I., 1991. Sustainable development: The gender perspective. *Development* 34, 134–135.

De-Luis, P., Bernal, E., Vela, M.J., Pérez, M., 2010. Stakeholders, board's gender balance and environment in corporate responsibility; pursuing sustainability. *International Business Management* 4, 229–238.

Dür, M., Keller, L., 2018. Education for sustainable development through international collaboration. A case study on concepts and conceptual change of school-students from India and Austria on gender equality and sustainable growth. *Education Sciences* 8, 1–17.

Franz-Balsen, A., 2010. Gender Mainstreaming and Mainstreaming Sustainability – Pathfinder for a New Science Approach. *Gaia* 19, 155–157.

Fudge, J., 2011. Gender, equality and capabilities: Care work and sustainable development. In Novitz T., Mangan D. (Eds.) The role of labour standards in development: From theory to sustainable practice? Oxford University Press. p. 40–57.

Furlotti, K., Mazza, T., Tibiletti, V., Triani, S., 2019. Women in top positions on boards of directors: Gender policies disclosed in Italian sustainability reporting. *Corporate Social Responsibility and Environmental Management* 26, 57–70.

Galbreath, J., 2011. Are there gender-related influences on corporate sustainability? A study of women on boards of directors. *Journal of Management and Organization* 17, 017–038.

Galmiche-Tejeda, A., Townsend, J., 2006. Sustainable development and gender hierarchies: Extension for semi-subsistence fish farming in Tabasco, Mexico. *Gender, Technology and Development* 10, 101–126.

Gligorijević, V., Vojković, G., Mirić, N., Vasić, P., 2016. Regional and gender differences in self-employment motivation: Implications on job sustainability. Mitteilungen der Osterreichischen Geographischen Gesellschaft 158, 149–166.

Goetz, A.M., Jenkins, R., 2016. Gender, security, and governance: The case of Sustainable Development Goal 16. *Gender and Development* 24, 127–137.

Graef, F., Hernandez, L.E.A., König, H.J., Uckert, G., Mnimbo, M.T., 2018. Systemising gender integration with rural stakeholders' sustainability impact assessments: A case study with three low-input upgrading strategies. *Environmental Impact Assessment Review* 68, 81–89.

Grosser, K., 2009. Corporate social responsibility and gender equality: Women as stakeholders and the European Union sustainability strategy. *Business Ethics-a European Review* 18, 290–307.

Hanson, S., 2010. Gender and mobility: New approaches for informing sustainability. *Gender, Place and Culture* 17, 5–23.

Huang, G., Ali, S., 2015. Local sustainability and gender ratio: Evaluating the impacts of mining and tourism on sustainable development in Yunnan, China. *International Journal of Environmental Research and Public Health* 12, 927–939.

Jacobson, J.L., 1992. Gender bias: Roadblock to sustainable development. Worldwatch Paper 110. Retrieved at 7th December 2019 from www.scopus.com/inward/record.uri?eid=2-s2.0-0027064184&partnerID=40&md5=65349a2f6ffed4c435e2f4c039ea8c6c

Kassinis, G., Panayiotou, A., Dimou, A., Katsifaraki, G., 2016. Gender and Environmental Sustainability: A Longitudinal Analysis. *Corporate Social Responsibility and Environmental Management* 23, 399–412.

Koehler, G., 2016. Tapping the Sustainable Development Goals for progressive gender equity and equality policy? *Gender and Development* 24, 53–68.

Lehtonen, M., 2004. The environmental–social interface of sustainable development: Capabilities, social capital, institutions, *Ecological Economics* 49 (2), 199–214.

Lisboa, T.K., Lusa, M.G., 2010. Sustainable development from a gender perspective – Brazil, Mexico, and Cuba: Women as protagonists in rural areas. *Revista Estudos Feministas* 18, 871–887.

Manandhar, M., Hawkes, S., Buse, K., Nosrati, E., Magar, V., 2018. Gender, health and the 2030 Agenda for sustainable development. *Bulletin of the World Health Organization* 96, 644–653.

Manjengwa, J.M., 1998. Environmental education for sustainable development in secondary schools in Zimbabwe: A focus on gender differences. *Environmental Education and Information* 17, 17–26.

Mapparessa, N., Bakry, M.I., Totanan, C., Mile, Y., Arumsari, A., 2017. The effect of political visibility, company characteristics and gender diversity to sustainability report disclosure. *International Journal of Civil Engineering and Technology* 8, 1019–1028.

Mauerhofer, V., 2008. 3-D Sustainability: An approach for priority setting in situation of conflicting interests towards a Sustainable Development. *Ecological Economics* 64, 496–506.

Mauerhofer, V., 2012. A 'Legislation-Check' based on '3-D Sustainability' – Addressing global precautionary land governance. *Land Use Policy* 29, 652–660.

Mauerhofer, V., 2013. Social capital, social capacity and social carrying capacity: Perspectives for the social basics within environmental sustainability. *Futures* 53, 63–73.

Mauerhofer, V., 2019. Legal institutions and ecological economics: Their common contribution for achieving a sustainable development. *Ecological Economics* 156, 350–359.

McClanahan, T.R., Abunge, C., 2017. Fish trader's gender and niches in a declining coral reef fishery: Implications for sustainability. *Ecosystem Health and Sustainability* 3 (6), 1–21.

Meek, W.R., Sullivan, D.M., 2018. The influence of gender, self-identity and organizational tenure on environmental sustainability. *Journal of Developmental Entrepreneurship* 23 (03), 1850018.

Meinzen-Dick, R., Kovarik, C., Quisumbing, A.R., 2014. *Gender and sustainability. Annual Review of Environment and Resources*, pp. 29–55.

Miotto, G., López, M.P., Rodríguez, J.R., 2019. Gender equality and UN sustainable development goals: Priorities and correlations in the top business schools' communication and legitimation strategies. Sustainability (Switzerland) 11.

Miralles-Guasch, C., Martinez Melo, M., Marquet, O., 2016. A gender analysis of everyday mobility in urban and rural territories: From challenges to sustainability. *Gender Place and Culture* 23, 398–417.

Mwita, E.J., Murphy, S.P., 2017. Challenging hidden hegemonies: Exploring the links between education, gender justice, and sustainable development practice. *Ethics and Social Welfare* 11, 149–162.

Nadeem, M., Zaman, R., Saleem, I., 2017. Boardroom gender diversity and corporate sustainability practices: Evidence from Australian Securities Exchange listed firms. *Journal of Cleaner Production* 149, 874–885.

Najjar, D., Spaling, H., Sinclair, A.J., 2013. Learning about sustainability and gender through Farmer Field Schools in the Taita Hills, Kenya. *International Journal of Educational Development* 33, 466–475.

Olsson, D., Gericke, N., 2017. The effect of gender on students' sustainability consciousness: A nationwide Swedish study. *Journal of Environmental Education* 48, 357–370.

O'Manique, C., Fourie, P., 2016. Affirming our world: Gender justice, social reproduction, and the sustainable development goals. *Development (Basingstoke)* 59, 121–126.

Oparaocha, S., Dutta, S., 2011. Gender and energy for sustainable development. *Current Opinion in Environmental Sustainability* 3, 265–271.

Palmer-Jones, R., Jackson, C., 1997. Work intensity, gender and sustainable development. *Food Policy* 22, 39–62.

Parida, P.C., Sinha, A., 2010. Performance and sustainability of self-help groups in India: A gender perspective. *Asian Development Review* 27, 80–103.

Rawls, J., 1971. *A Theory of Justice*. Harvard University Press, Cambridge.

Remington-Doucette, S., Musgrove, S., 2015. Variation in sustainability competency development according to age, gender, and disciplinary affiliation: Implications for teaching practice and overall program structure. *International Journal of Sustainability in Higher Education* 16, 537–575.

Rozario, S., Samuel, G., 2010. Gender, religious change and sustainability in Bangladesh. *Womens Studies International Forum* 33, 354–364.

Saey-Volckrick, J., 2020. What does the rebound effect tell us? Reflection on its sources and its implication for the sustainability debate. In: Mauerhofer V, Rupo D, Tarquinio L (Eds.), *Sustainability and Law*, Cham: Springer pp. 103–118.

Shoham, A., Almor, T., Lee, S.M., Ahammad, M.F., 2017. Encouraging environmental sustainability through gender: A micro-foundational approach using linguistic gender marking. *Journal of Organizational Behavior* 38, 1356–1379.

Slepian, J.L., Jones, G.E., 2013. Gender and corporate sustainability: On values, vision, and voice. *Organisation Management Journal* 10, 215–226.

Smetschka, B., Gaube, V., Lutz, J., 2009. Integrating a gender perspective into the sustainability triangle via time-use analysis. *Journal of the Austrian Society of Agricultural Economics* 18, 135–148.

Soares, D., 2006. Gender, firewood and sustainability: The case of a community in Los Altos de Chiapas. *Economia Sociedad Y Territorio* 6, 151–175.

Thomas-Slayter, B.P., Rocheleau, D.E., 1994. Essential connections: Linking gender to effective natural resource management and sustainable development. Working Paper – Michigan State University, Office of Women in International Development 242.

Tisdell, C., 2001. Sustainable development, gender inequality and human resource capital. *International Journal of Agricultural Resources, Governance and Ecology* 1, 178–192.

Turner, J., 2012. Urban mass transit, gender planning protocols and social sustainability – The case of Jakarta. *Research in Transportation Economics* 34, 48–53.

UN (2012). Resolution adopted by the General Assembly on 27 July 2012, "The future we want" A/RES/66/288, Retrieved at 27th September 2016 from www.un.org/ga/search/view_doc.asp?symbol=A/RES/66/288&Lang=E

UN (2015). Resolution adopted by the General Assembly on 25 Sept 2015, "Transforming our world: The 2030 Agenda for Sustainable Development" A/RES/70/1, Retrieved at 27th September 2016 from www.un.org/en/development/desa/population/migration/generalassembly/docs/globalcompact/A_RES_70_1_E.pdf

Valls Martínez, M.D.C., Cruz Rambaud, S., Parra Oller, I.M., 2019. Gender policies on board of directors and sustainable development. *Corporate Social Responsibility and Environmental Management* 26, 1539–1553.

Vu, H.A., 2012. Gender, sustainability, and shrimp farming: Negotiating risky business in Vietnam's Mekong Delta. Cruz-Torres M. L., McElwee P. (Eds.) *Gender and Sustainability: Lessons from Asia and Latin America*. University of Arizona Press, pp. 165–186.

Vyas-Doorgapersad, S., 2014. Gender, ICT and millennium development goals for sustainable development in West Africa. *Mediterranean Journal of Social Sciences* 5, 405–412.

WCED, 1997, Our Common Future, Report of the World Commission on Environment and Development, Oxford University Press. 383 pp.

Weaven, S., Griffin, D., McPhail, R., Smith, C., 2013. Who is listening? An examination of gender effects and employment choice in sustainability education in an undergraduate business school. *Australian Educational Researcher* 40, 567–582.

Wickramasinghe, A., 2003. Gender and health issues in the biomass energy cycle: Impediments to sustainable development. *Energy for Sustainable Development* 7, 51–61.

Wiium, N., Wreder, L.F., Chen, B.B., Dimitrova, R., 2019. Gender and positive youth development: Advancing sustainable development goals in Ghana. *Zeitschrift fur Psychologie / Journal of Psychology* 227, 134–138.

Zahid, M., Rahman, H.U., Ali, W., Khan, M., Alharthi, M., Imran Qureshi, M., Jan, A., 2020. Boardroom gender diversity: Implications for corporate sustainability disclosures in Malaysia. *Journal of Cleaner Production* 244, 118683.

Zamora, G., Koller, T.S., Thomas, R., Manandhar, M., Lustigova, E., Diop, A., Magar, V., 2018. Tools and approaches to operationalize the commitment to equity, gender and human rights: Towards leaving no one behind in the Sustainable Development Goals. *Global Health Action* 11, 1–7. doi: 10.1080/16549716.2018.1463657

Zein-Elabdin, E., 1997. Gender equity and sustainable development in sub-Saharan Africa: The World Bank strategy for the forest sector. *Scandinavian Journal of Development Alternatives* 16, 75–95.

4

GENDER MAINSTREAMING IN OCCUPATIONAL HEALTH AND SAFETY

Challenges in the era of the Sustainable Development Goals

Hiroaki Matsuura

4.1 Introduction

Over the past 100 years, the proportion of women in the workforce has increased dramatically (Killingsworth and Heckman, 1986; ILO, 2021b). They now make up 38.9% of the workforce worldwide (ILO, 2021b). Thanks to this trend, a larger number of women find themselves contending with a wide variety of occupational risk factors. However, women still represent only a small fraction of the total burden of disease as caused by occupational risk factors. According to the Global Burden of Disease Study 2019, occupational risk factors accounted for 269,010 deaths and 3.7% of global disability-adjusted life years (DALYs) among women, compared to 951,491 deaths and 6.7% of global DALYs among men worldwide. Women still account for only 22% of total deaths and 29% of global DALYs caused by occupational risk factors worldwide (Murray et al., 2020).

Due to the segmentation of the genders in specific occupations and sectors, generally speaking, women tend to be exposed to lower occupational health risks, leading to a lower burden of disease and injury. However, there are several concerns over the interpretation of the statistics. First, women are more likely than men to engage in agricultural work, domestic work, and other forms of informal work (ILO, 2021b). This work often remains unrecognized or uncounted in the official statistics and is less likely to be placed under the protection of occupational health and safety legislations. For this reason, the burden of occupational health for women might be underestimated. Second, the perception that women's work is safer and that any health problems identified among female workers in male-dominant workplaces can be attributed to unfitness for the working environment leads to underreporting of injuries and accidents among female workers. It is reported that when female workers do experience a work-related health problem, they are less likely to be diagnosed, recognized, compensated, or offered rehabilitation, compared

DOI: 10.4324/9781003174257-5

to their male counterparts (WHO, 2011a). Third, women have been historically underrepresented in many sectors of the formal economy. This creates a situation in which current health and safety standards and policies are set based on the scientific evidence using male-dominated data from the past (EU-OSHA, 2014). Women may be, in fact, at a greater risk than men if such health and safety standards are sufficient to protect men, but not women.

Women and men are different in terms of not only their biological characteristics but also their jobs, working conditions, circumstances, and how the society treats them. All these factors can potentially affect the risks they face in the workplace, and thus policy measures must be taken to prevent these risks. Adopting a gendered perspective when thinking about occupational health and safety legislation is a relatively recent phenomenon. The majority of countries still do not have any occupational health and safety legislations focusing on the gender differences in occupational health and safety with the exception of specific protection for pregnant women (ILO, 2014). However, there is a growing consensus that a "one-size-fits-all" approach to occupational health and safety regulation, ignoring issues of gender differences, is harmful to women's health and well-being in the workplace. Recognizing the heterogeneity of workers in the workforce is vital in ensuring the safety and health of all workers regardless of gender and other differences.

This chapter discusses the challenges of gender mainstreaming in the field of occupational health and safety in the era of the Sustainable Development Goals (SDGs). Understanding the sources of gender differences in occupational health and advancing gender mainstreaming in this field can contribute to the achievement of Goal 3 (good health and well-being), Goal 5 (gender equality), Goal 8 (decent work and economic growth), and Goal 16 (peace, justice, and strong institutions) of the SDGs. Sections 4.2 to 4.5 lay out the potential sources of gender differences in occupational health. More specifically, Section 4.2 compares the differences in exposure to occupational health hazards between men and women due to the different types of occupational segregation. Section 4.3 discusses the gender differences in the heterogeneous effect of the same chemical, physical, and psychological hazards in the workplace. Section 4.4 examines the protection of women's reproductive health in the workplace, which remains a fixture in many countries' occupational health and safety legislations. Section 4.5 explores the new challenges related to women's workplace health and safety under the current COVID-19 pandemic. Section 4.6 looks at the progress of gender mainstreaming in this area at the international and national level. Finally, Section 4.7 concludes this chapter and considers the future direction of gender mainstreaming in the post COVID-19 pandemic world.

4.2 Unequal distribution of jobs positions and tasks

Gender-based labor market segregation contributes to different levels of exposure to occupational health and safety risks for women and men. The pattern of women's employment is very different from country to country, but "horizontal segregation," which refers to under- or over-representation of women in particular industries, is persistent across all countries. "Vertical segregation" is also common to all industries across all countries, as women are underrepresented in positions of authority and have less job security, lower control of their working environment, worse contractual working conditions, and lower job status within the same sector (Campos-Serna et al., 2013).

Due to the ideology of the gender-based division of labor, household tasks and family care are typically assigned to women, and even working women in many countries are not exempted from these expectations. Balancing the demands of paid work in the labor market and unpaid household tasks and care work, female workers often need to take low or non-paying "flexible" jobs with little or no protection.

In developing countries, women contribute to their communities mainly through agricultural work or informal employment in which case their work is invisible from public scrutiny and may not be recognized as "real" work even by the women themselves (ILO, 2021b). Along with construction and mining, the agriculture sector consistently ranks among the most dangerous occupational industries, bearing a high risk of job-related injuries, illnesses, and death (ILO, 2021b). Agricultural workers face a wide variety of health risks from injury from agricultural machinery to infection from biological hazards. Exposure to pesticides is one of the major occupational risks for all agricultural workers, but female agricultural workers are exposed to more pesticides than their male counterparts, because they are responsible for washing pesticide-soaked clothing and are less likely to read warning labels due to their low literacy (WHO, 2011a).

According to the most recent estimates, employment in the informal economy represents about 80% of female workers in the non-agricultural sector in South Asia and Sub-Saharan Africa (Razavi and Turquet, 2016). Women in the informal economy are more often found in vulnerable situations than their male counterparts, including domestic workers, home-based workers, unpaid family workers, and sex workers.

Home workers, for example, face occupational health risks including overworking, ergonomic problems, mental health disorders, injury, and exposure to harmful chemicals (ILO, 2021c). Their workplace (home) rarely satisfies health and safety standards set by governmental agencies, because no labor inspection is expected at private dwellings. Domestic workers also face a variety of workplace hazards when working in clients' homes, including exposure to harmful cleaning chemicals, verbal and physical abuse, and injuries caused by lifting and moving clients with limited mobility (Smith, 2011). Sex workers are disproportionately exposed to sexually transmitted diseases such as HIV and syphilis (Baral et al., 2012; WHO, 2021). Occupational injuries and accidents among women working in the informal economy are also less likely to be counted in occupational health and safety statistics, which means their occupational health risks are severely underestimated.

In the EU and the United States, where women are more likely to engage in formal-sector activities, women are still less likely to find employment in sectors such as construction, mining, agriculture, and fishery, where workers are more exposed to risks such as falls, explosions, electrocutions, chemicals, and noise. On the other hand, they are more likely to work in low-wage manufacturing, retail, health, and education-related sectors where workers are more likely to be exposed to occupational risks such as biological and chemical hazards, mental health problems (as a result of monotonic work, high demands, and limited authority), and musculoskeletal disorders and cardiovascular diseases (as a result of repetitive movements and awkward static and standing postures) (WHO, 2011a). Women are also at risk of workplace violence because they are over-represented in jobs where they must interact with pupils, clients, and patients in retail, health, and education-related sectors (Lippel, 2001; Hogh et al., 2008; WHO, 2011a).

Recent research shows that even after adjusting for work characteristics, women are still subject to more than twice the risk of workplace violence than men (Lanthier et al., 2018). This may be due to the third type of occupational segregation, namely "task segregation." Unlike horizontal and vertical segregation, "task segregation" is an almost invisible and completely neglected aspect in the literature of occupational health and safety. For example, male cleaners are more likely to mop, while female cleaners are more likely to dust (Messing et al., 1998). Because of the existence of task segregation, women and men with the same job titles in the same occupations can still have very different rates of exposure to occupational risks, which makes the quantitative studies of occupational health and safety more difficult to interpret.

4.3 Lack of evidence based on female workers' data in male-dominated workplace

Since a large amount of the evidence in relation to workers' data from the 20th century is taken from male-dominated workplaces, it is questionable whether the existing scientific evidence is applicable to the growing female working population. Most epidemiological research studies have either ignored or merely controlled for gender in their analyses. The limited number of studies have investigated the heterogeneous effect of the same chemical, physical, and psychological hazards by gender in the workplace.

For chemical hazards, susceptibility to chemicals varies depending on how quickly and efficiently a toxic agent is metabolized; it is therefore affected by the four phases of the toxicokinetic process: absorption across the body barriers, the distribution into the body, the metabolism, and the excretion (Scarselli et al., 2018). Scarselli and his colleagues also identified some of the factors for the gender differences in each phase of the toxicokinetic process: (1) condition of skins, number of hair follicles, breathing rates and respiratory volume, and gastrointestinal motility; (2) body fat composition, weight, plasma volume, and average organ blood flow; (3) difference in xenobiotic metabolic activity, which is affected by gene expression for enzymes of the CYP450 superfamily; and (4) rate of renal excretion of compounds. Other studies showed that sensitivity to asthmatic attacks increases in the premenstrual phase and airway reactivity to allergens and irritants varies over time and with hormones (Tollefsen et al., 2007).

For physical hazards, on average, women are at a higher risk than men for work-related injuries, particularly those with musculoskeletal symptoms (MacRae, 2005). Many ergonomic designs for work equipment, tools, and personal protective equipment are still based on average male physical capacity (ILO, 2013). As a result, not only women but also men who have difficulty finding suitably sized equipment feel uncomfortable with such designs and face a higher risk of musculoskeletal disorders.

The heterogeneous effect between sexes also comes from cultural factors. As we discussed in the previous section, household tasks and family care are typically assigned to women, and even working women are not exempted from these expectations. Female workers are more likely to experience long workdays as a result of combining paid and unpaid work. Unfortunately, only working hours for paid work are regulated by labor law. Therefore, equal workload between men and women may lead to different levels of stress, mental health, and musculoskeletal disorders between the sexes.

As was also discussed in the previous section, women are more likely to face workplace violence. While it is plausible, there is no existing study to support the notion that women are more susceptible than men to developing mental distress when exposed to the same workplace violence on an individual level. At a more aggregated level, however, a few studies have investigated gender differences in the effects of a violent working environment on workers' mental health risks (Pien et al., 2015).

4.4 Sexual and reproductive health

As many women work during their reproductive years, they are exposed to a wide variety of chemical, physical, and psychological hazards in the workplace that can potentially cause problems such as infertility, miscarriage, birth defects, and developmental disorders in children. Not surprisingly, the epidemiological research on the reproductive effects of occupational hazards has been intensively studied among women (Walsh and Kelleher, 1987).

In the past, women's opportunities for employment were significantly hindered by overprotective occupational health and safety regulations. All women, not just pregnant women, were severely restricted from working overtime, late at night, or in dangerous industries such as mining. Over time, these restrictions have been removed, but protection for pregnant women still remains in many OECD countries' legislation since reproduction is viewed as women's domain. However, the focus of the research has recently expanded to include male-mediated reproductive effects (Goldman et al., 2013). The male reproductive system may be affected by workplace hazards through lowered sperm count, abnormal sperm shape, altered sperm transfer, altered sexual performance, and adverse pregnancy outcomes

(WHO, 2011a). Some of these researches may be translated into the future occupational health and safety regulations.

Menstrual disorders can affect the work attendance and performance of female workers, but this has long been neglected in the literature of occupational health. Among all, dysmenorrhoea has been cited as one of the most common reasons for periodic or short-term sick leave from work and school among young women (Weissman et al., 2004). Japan was the first country to introduce menstrual leave into its national labor laws. Article 68 of the Labour Standards Law of 1947 states "When a woman for whom work during menstrual periods would be especially difficult has requested leave, the employer shall not employ such woman on days of the menstrual period." To date, menstrual leave policies exist in a limited number of other countries such as Taiwan, China, South Korea, Indonesia, Zambia, and Mexico (Levitt and Barnack-Tavlaris, 2020). In 2017, Italy considered introducing a menstrual leave policy into its labor laws, but the motion failed to secure enough votes. There is a concern that menstrual leave may lead to unintended consequences, such as discrimination in hiring practices, salaries or wages, and promotion, but there is no convincing empirical evidence to date that establishes this causal effect (Herrmann and Rockoff, 2012; King, 2021).

4.5 Gender and occupational health under the COVID-19 pandemic

The pandemic has disproportionately affected female workers worldwide, with many women working on the frontline in the fight against COVID-19. The report from the OECD correctly addressed that women are "over-represented working in health systems, continue to do the majority of unpaid care work in households, face high risks of economic insecurity, and face increased risks of violence, exploitation, abuse or harassment during times of crisis and quarantine" (OECD, 2020).

The COVID-19 pandemic has had a pronounced negative impact on women's employment, because women tend to work in the sectors most affected by the spread of the virus. They work in service sectors that require physical interaction with customers, including air travel, tourism, retail activities, accommodation services, and food and beverage service activities. Women make up approximately 47% of workers in the air transport industry, 53% in food and beverage services, and 60% in accommodation services (ILO, 2021b). In the retail sector, on average, 62% of workers are women (ILO, 2020; ILO, 2021b).

Globally, 70% of all health workers are women, and they have faced many stressful situations at work during the pandemic (ILO, 2020). Increased workloads, longer working hours, and reduced rest periods have become the norm during the crisis (OECD, 2020).

Remote work has been widely adopted worldwide to maintain job continuity and prevent the spread of COVID-19 at the same time. Before the pandemic, remote work, or telework, was considered to be one of the most promising strategies for achieving a "work–life balance" through flexible working arrangements for carers and parents of young children in the European Union (EC, 2020). However, many workers who opted for this style of work struggled with the same problems experienced by home workers, such as work intensification and excessive hours, the blurring of boundaries between paid work and private life, musculoskeletal disorders, the responsibility of caring for children or parents while working, and social isolation (ILO, 2021c). The number of cases of domestic violence also rose and workers confined to their homes due to lockdowns also faced such violence (UN Women, 2020).

In summary, the COVID-19 pandemic has affected both men and women, but more specifically has exaggerated two existing problems in women's working environment. These include lower job security, lower control of their working environment, and lower job status within the same sector, as well as occupational health and safety problems specific to home workers and domestic workers.

4.6 Policy response: Mainstreaming gender in occupational health and safety

While Article 11 of the Convention on the Elimination of All Forms of Discrimination against Women of 1979 guaranteed the equal protection of occupational health and safety with special protection granted to women's reproductive function, taking gender differences into account for other areas of occupational health and safety is a relatively recent phenomenon (UNGA, 1979). Over recent years, the policy focus of gender issues in occupational health and safety has shifted from the special protection of women based on an assumed difference between women and men to the equalization of rights and protection for all workers regardless of their gender. However, as we discussed in the previous section, the gendered nature of occupational health has remained in place until now. Women's health issues in the workplace are still overlooked or minimized in many parts of the world. "Gender mainstreaming" requires assessing from the start the implications for women and men of any policy action including legislation, policies, or programs in any area and at all levels (UNDESA, 1997). It shifted from an exclusive focus on women to a focus on gender, ensuring that both men and women benefit equally from planned policy action.

At the international level, the Fourth World Conference on Women in 1995 endorsed gender mainstreaming as a critical and strategic approach for achieving gender equality. The Beijing Declaration and Platform for Action, adopted in the conference, explicitly recognized the importance of tackling occupational health and safety issues to achieve equitable outcome between men and women (UN, 1995).

Two UN organizations, the International Labor Organization (ILO) and the World Health Organization (WHO), have primary roles in improving occupational health and safety worldwide. For more than 70 years, the two organizations have worked closely together in this area[1] (Joint ILO/WHO Committee on Occupational Health, 1992). Over time, there has been a shift from a labor law approach to a public health approach to occupational health in ILO. In the traditional labor law approach, employers have a responsibility to protect workers' health and safety in the workplace as per the terms of their labor contract. However, the public health approach promotes the health of all workers even beyond the workplace. This is reflected in the common definition of occupational health, which was adopted by the Joint ILO/WHO Committee on Occupational Health at its first session (1950) and revised at its twelfth session (1995), reading:

> Occupational health should aim at: the promotion and maintenance of the highest degree of physical, mental and social wellbeing of workers in all occupations; the prevention amongst workers of departures from health caused by their working conditions; the protection of workers in their employment from risks resulting from factors adverse to health; the placing and maintenance of the worker in an occupational environment adapted to his physiological and psychological capabilities; and, to summarize, the adaptation of work to man and of each man to his job.

For the ILO, occupational health and well-being are the crucial components of their decent work agenda. In 2001, the ILO asserted that "the best way to avoid a life of poverty is to find decent work" (ILO, 2001). Since then, the ILO has intensified its efforts to secure decent work for all workers and adopted the Global Strategy on Occupational Health and Safety in 2003 and the Promotional Framework for Occupational Health and Safety Convention and Recommendation (C187 and R197) in 2006. Article 5 of the 2006 Promotional Framework for Occupational Health and Safety Recommendation (R197) explicitly stated that "Members should take measures to protect the safety and health of workers of both genders, including the protection of their reproductive health." The 1996 Home Work Convention (No. 177) and the 2011 Domestic Workers Convention (No. 189) ensured that occupational health and safety legislations must apply to home domestic workers who were outside the scope of such laws before these conventions.

In 2013, ILO published "10 Keys for gender sensitive OSH practice – Guidelines for gender mainstreaming in occupational health and safety," explaining how gender issues can be integrated into analyses, formulation, and monitoring of policies, programs, and preventive measures in order to reduce inequalities between men and women (ILO, 2013). They include specific advice in the areas of information gathering, standard setting, legislation, policymaking, and preventive strategies. The latest report, "Empowering Women at Work – Government Laws and Policies for Gender Equality," published in 2021, affirmed that gender perspective is mainstreamed not only in ILO conventions and recommendations set out in international labor standards for gender equality but also in other ILO instruments on working time, part-time work, home-based work, social security, occupational health and safety, and domestic work (ILO, 2021a).

The WHO has also committed itself to the improvement of occupational health and safety. A proposal for a "Global Strategy for Occupational Health for All" was first discussed in the meeting of the WHO Collaborating Centers in 1994 and adopted by the World Health Assembly two years later (WHO, 1995). In 2007, the Global Plan of Action on Workers' Health 2008–2017 followed up this global strategy and set forth a 10-year agenda for a global response to the specific health needs of workers and urged Member States to work toward coverage of all workers, including those who work in the agricultural and informal sector as well as migrants, with essential interventions and basic health services for the improvement of occupational health (WHO, 2013). Over the past two decades, social determinants of health have been identified as key factors improving population health outcomes. In 2008, the WHO's Commission on Social Determinants of Health published "Closing the Gap in a Generation: Health Equity through Action on the Social Determinants of Health" (WHO, 2008). The Commission classed fair employment/decent work as one of the key strategies for narrowing health inequality, along with giving every child the best start in life, developing healthy communities and environments, and establishing adequate social and health protection. The report assured that occupational health and safety programs would be applied to all workers, both formal and informal, and provided a more integrated view of population health across workplace and non-workplace domains over the life course. The report also recommended the use of gender mainstreaming as a key strategy for achieving gender equality and ultimately reducing health gaps. The WHO Department of Gender, Women and Health developed a monitoring and evaluation framework in order to confirm their commitment to gender mainstreaming at all levels of their work (WHO, 2011b).

The 2030 Agenda for Sustainable Development, adopted by the United Nations General Assembly in 2015, calls on Member States to begin their efforts to achieve the 17 Goals and 169 targets over the next 15 years (UNGA, 2015). Occupational health is related to the "Good Health and Wellbeing" SDG, namely its target 3.9: "by 2030 substantially reduce the number of deaths and illnesses from hazardous chemicals and air, water, and soil pollution and contamination"; the "Decent Work and Economic Growth" SDG, namely its target 8.8: "protect labor rights and promote safe and secure working environments of all workers, including migrant workers, particularly women migrants, and those in precarious employment"; and the "Peace, Justice and Strong Institutions" SDG, namely its target 16.6: "develop effective, accountable and transparent institutions at all levels" (ILO, 2019). Moreover, many targets in the "Gender Equality" SDG, including Targets 5.1, 5.2, 5.4, 5.5, 5.6, 5A, 5B, and 5C, provide both goals and means of gender mainstreaming in occupational health and safety. Both ILO and WHO are working toward these targets with other relevant stakeholders.

In addition to these two international organizations, the European Union (EU) and its Member States are also interested in gender mainstreaming in occupational health and safety. Traditionally, the approach of EU regulations and directives on occupational health and safety has been "gender neutral" with the exception of Council Directive 92/85/EEC, which refers to the protection of pregnant and breastfeeding female workers (EU-OSHA, 2014). The European Agency for Safety and Health at Work (EU-OSHA) was not an exception and used a male-dominated sample and simply generalized the findings to women working in male-dominant sectors under the name of a "gender neutral" policy (EU-OSHA, 2014).

The introduction of "Community Strategy on Safety and Health at Work 2002–06" formally incorporated gender analysis into occupational health and safety practice in the EU. After another five-year period, 2007–2012, the European Commission successfully identified the common objectives and priorities among Member States and adopted a new **EU Strategic Framework on Health and Safety at Work 2014–2020**. The Strategic Framework includes gender mainstreaming as an objective and supports EU-OSHA to carry out research examining gender differences in risk assessment, preventive measures, and compensation arrangements (EC, 2017). The EU Strategic Framework on Health and Safety at Work 2014–2020 highlighted that women can face specific risks and that occupational health policy can contribute to promoting equal opportunities (Flaspöler et al., 2013; EU-OSHA, 2014). Furthermore, in 2019, the EU adopted a new Directive on Work-Life Balance for Parents and Carers, which aims to increase labor market participation of women through family-related leave and flexible working arrangements (EC, 2020). A new Directive requires Member States to provide flexible working arrangements for carers and parents of young children through the use of remote working arrangements, flexible working schedules, or reduced working hours by 2022.

In Canada, the Canadian Centre for Occupational Health and Safety (CCOHS) is a federal agency that promotes the health and well-being of Canadian workers by providing information, education, and solutions for workplaces. In 2011, Gender-based Analysis Plus (GBA+) was introduced as a formal federal policy tool to assess the different experiences of women and men of policies, programs, and initiatives of the Canadian Government. In the 2019–2020 Departmental Results Report, CCOHS explicitly recognized the importance of GBA+ and committed to ensuring diverse groups of workers in Canada benefit from its programs (Minister of Labor, 2020). In 2017, CCOHS launched the "Gender, Work and Health" web portal, which aims to bridge the gap between gender, sex, and health, and their impact on the workplace.

Outside of the EU member states and Canada, gender mainstreaming has not yet become an integral part of occupational health and safety regulations even in other G7 countries.

In the United States, the National Institute for Occupational Health and Safety (NIOSH) is a federal research agency that has focused on the study of worker safety and health since 1970. Gender and occupational health have long been a research area for the institute. Several reports have been published in relation to this theme[2] (Kasner et al., 2012). However, it is an internal policy of the NIOSH rather than a statutory requirement or strategic framework of federal government occupational health and safety regulations.

In Japan, the National Institute of Occupational Safety and Health (JNIOSH) is the governmental research institute for occupational safety and health. The gender mainstreaming in occupational health research and policy process is extremely slow at the national level. In 2021, the Japan Society for Occupational Health Policy and Legislation Committee published "Proposal: Supporting the Health of Female workers," which was one of the first documents to commit to gender mainstreaming in the area of occupational health and safety (Japan Society for Occupational Health, 2021).

4.7 Discussion

Women and men are different in terms of not only their biological characteristics but also their jobs, working conditions, and circumstances and how the society treats them. All these factors can affect the risks they face in the workplace and also the approach taken to prevent these potential risks. In the past 25 years, there has been modest progress in gender mainstreaming in the field of occupational health and safety at both the international and national level. However, the progress of gender mainstreaming remains limited, especially outside of the EU Member States and Canada.

In order to make the workplace accessible to both women and men with no discrimination, all stakeholders must take into account diversity among all workers related to gender differences. Such a process requires the following: identifying gaps in gender equality in health risks, exposures, and outcomes through the use of gender-disaggregated data; developing strategies to close those gaps through research, public policy, and in the workplace; putting resources and expertise into implementing such strategies; monitoring and evaluation of implementation; and holding institutions accountable for policy outcomes. When women are in disadvantaged position, the use of women-specific activities or affirmative action might be an option. However, we must ensure that such activities are not used erroneously to justify occupational segregation or inadequate occupational health and safety interventions.

The COVID-19 pandemic introduced new challenges to gender issues in occupational health and safety. However, little attention is paid to gender differences in terms of the exposure to and the impact of occupational health risks newly created by the pandemic and containment measures. The use of remote work, which was once considered to be a promising approach to achieving a work–life balance in the EU, has created additional occupational health challenges. Such negative health consequences would be at least partly mitigated if employers or contracted inspectors provided adequate safety interventions in workers' private dwellings. In France, "the right to disconnect" was introduced in an amendment to the Labor Code in 2016, allowing workers to disconnect from email and other communication technology after working hours (ILO, 2021c). Such an arrangement is also helpful for preventing remote workers from overworking. The special needs of pregnant women should be considered, as should the needs of other workers who are particularly vulnerable to COVID-19 and other occupational risks.

There are many old and new challenges in this area, but occupational health and safety has been well integrated into the SDGs. The SDGs are certainly not a quick fix to solve all occupational health and gender inequity problems, but rather a strong vehicle to create better, healthier, more transparent, and more gender-sensitive workplaces. More progress will be expected in this area in the next decade.

Notes

1 There is a significant overlap in their interest, but generally speaking, the focus of the ILO has been on the provision of international guidelines and a legal framework for the development of policies as well as practical support to improve actions at the workplace, while the WHO has concentrated more on the provision of scientific backgrounds, methodologies, technical support, and on the training of health and related manpower for occupational health.

2 Gender and Toxicity (1997), Female Workers Face High Risks From Work Stress, Musculoskeletal Injuries, Other Disorders (2000), Women's Safety and Health Issues at Work (2001), Gender Differences in Acute Pesticide-Related Illnesses and Injuries.

References

Baral, S., Beyrer, C., Muessig, K., Poteat, T., Wirtz, A. L., Decker, M. R., Sherman, S. G. & Kerrigan, D. 2012. Burden of HIV among female sex workers in low-income and middle-income countries: A systematic review and meta-analysis. *Lancet Infectious Diseases*, 12, 538–549.

Campos-Serna, J., Ronda-Perez, E., Artazcoz, L., Moen, B. & Benavides, F. 2013. Gender inequalities in occupational health related to the unequal distribution of working and employment conditions: A systematic review. *International Journal for Equity in Health*, 12, 57.

EC. 2017. *Safer and healthier work for all – Modernisation of the EU occupational safety and health legislation and policy* [Online]. Brussels: European Commission. Available: https://eur-lex.europa.eu/legal-content/EN/TXT/?uri=CELEX%3A52017DC0012 [Accessed 5/21/2021].

EC. 2020. Directive on work-life balance for parents and carers. In: Commission, E. (ed.) *Directive (EU) 2019/1158*. Brussels: European Commission.

EU-OSHA. 2014. *Mainstreaming gender into occupational safety and health practice* [Online]. European Union. Available: https://osha.europa.eu/en/publications/reports/6805688 [Accessed 5/21/2021].

Flaspöler, E., Hauke, A., Koppisch, D., Reinert, D., Koukoulaki, T., Vilkevicius, G., Žemės, L., Águila Martenez-Casariego, M., Baquero Martenez, M. & González Lozar, L. 2013. *New risks and trends in the safety and health of women at work*.

Goldman, M. B., Troisi, R. & Rexrode, K. M. 2013. *Women and health*. Amsterdam, Elsevier/Academic Press.

Herrmann, M. A. & Rockoff, J. E. 2012. Does menstruation explain gender gaps in work absenteeism? *The Journal of Human Resources*, 47, 493–508.

Hogh, A., Sharipova, M. & Borg, V. 2008. Incidence and recurrent work-related violence towards healthcare workers and subsequent health effects. A one-year follow-up study. *Scandinavian Journal of Public Health*, 36, 706–712.

ILO. 2001. *Poverty reduction and decent work in a globalizing world* [Online]. Geneva: International Labor Organization. Available: www.ilo.org/global/docs/WCMS_078848/lang--en/index.htm [Accessed 5/21/2021].

ILO. 2013. *10 keys for gender sensitive OSH practice – Guidelines for gender mainstreaming in occupational safety and health* [Online]. Geneva: International Labor Organization. Available: www.ilo.org/global/topics/safety-and-health-at-work/resources-library/publications/WCMS_324653/lang--en/index.htm [Accessed 5/21/2021].

ILO. 2014. *Global database on occupational safety and health legislation* [Online]. Geneva: Internaational Labor Organization. Available: www.ilo.org/safework/info/publications/WCMS_217849/lang--en/index.htm [Accessed 5/21/2021].

ILO. 2019. *Relevant SDG targets related to health and safety at the workplace* [Online]. Geneva: International Labor Organization. Available: www.ilo.org/global/topics/dw4sd/themes/osh/WCMS_558571/lang--en/index.htm [Accessed 5/21/2021].

ILO. 2020. *COVID-19 and the world of work: Impact and policy responses* [Online]. Geneva: International Labor Organization. Available: www.ilo.org/wcmsp5/groups/public/---dgreports/---dcomm/documents/briefingnote/wcms_738753.pdf [Accessed 5/21/2021].

ILO. 2021a. *Empowering women at work – Government laws and policies for gender equality* [Online]. Geneva: ILO. Available: www.ilo.org/empent/Publications/WCMS_773233/lang--en/index.htm [Accessed 5/21/2021].

ILO. 2021b. *ILOSTAT* [Online]. Geneva: ILO. Available: https://ilostat.ilo.org/ [Accessed 5/15/2021].

ILO. 2021c. *Working from home: From invisibility to decent work* [Online]. Geneva: International Labour Organization. Available: www.ilo.org/global/publications/books/forthcoming-publications/WCMS_765806/lang--en/index.htm [Accessed 5/21/2021].

Japan Society for Occupational Health. 2021. *Proposal: Supporting the health of working women in Japan* [Online]. Tokyo: Japan Society for Occupational Health,. Available: www.sanei.or.jp/images/contents/370/Proposal_HWW_Policies_and_Regulations_Committee.pdf [Accessed 5/21/2021].

Joint ILO/WHO Committee on Occupational Health. 1992. *Eleventh session, document* [Online]. Geneva: International Labor Organization. Available: https://ilo.primo.exlibrisgroup.com/discovery/fulldisplay/alma992857463402676/41ILO_INST:41ILO_V2 [Accessed 5/21/2021].

Kasner, E. J., Keralis, J. M., Mehler, L., Beckman, J., Bonnar-Prado, J., Lee, S. J., Diebolt-Brown, B., Mulay, P., Lackovic, M. & Waltz, J. 2012. Gender differences in acute pesticide-related illnesses and injuries among farmworkers in the United States, 1998–2007. *American Journal of Industrial Medicine*, 55, 571–583.

Killingsworth, M. R. & Heckman, J. J. 1986. Female labor supply: A survey. *Handbook of Labor Economics*, 1, 103–204.

King, S. 2021. Menstrual leave: Good intention, poor solution. In: HASSARD, J. & TORRES, L. D. (eds.) *Aligning Perspectives in Gender Mainstreaming: Gender, Health, Safety, and Wellbeing*. Cham: Springer International Publishing.

Lanthier, S., Bielecky, A. & Smith, P. M. 2018. Examining risk of workplace violence in Canada: A sex/gender-based analysis. *Annals of Work Exposures and Health*, 62, 1012–1020.

Levitt, R. B. & Barnack-Tavlaris, J. L. 2020. Addressing Menstruation in the Workplace: The Menstrual Leave Debate. In: BOBEL, C., WINKLER, I. T., FAHS, B., HASSON, K. A., KISSLING, E. A. & ROBERTS, T.-A. (eds.) *The Palgrave Handbook of Critical Menstruation Studies*. Singapore: Springer Singapore.

Lippel, K. 2001. Les agressions au travail: un même traitement pour les travailleurs et les travailleuses? *Recherches féministes*, 14, 83–108.

Macrae, N. 2005. Women and work: A ten year retrospective. *Work,* 24, 331–339.

Messing, K., Chatigny, C. & Courville, J. 1998. 'Light' and 'heavy' work in the housekeeping service of a hospital. *Applied Ergonomics*, 29, 451–459.

Minister of Labor. 2020. *2019–20 Departmental Results Report for the Canadian Centre for Occupational Health and Safety (CCOHS)* [Online]. Hamilton: Minister of Labor,. Available: www.ccohs.ca/ccohs/reports/dpr/2019-20.html [Accessed 5/21/2021].

Murray, C. J., Aravkin, A. Y., Zheng, P., Abbafati, C., Abbas, K. M., Abbasi-Kangevari, M., Abd-Allah, F., Abdelalim, A., Abdollahi, M. & Abdollahpour, I. 2020. Global burden of 87 risk factors in 204 countries and territories, 1990–2019: A systematic analysis for the Global Burden of Disease Study 2019. *The Lancet,* 396, 1223–1249.

OECD. 2020. *Women at the core of the fight against COVID-19 crisis* [Online]. Paris: OECD. Available: https://books.google.co.jp/books?id=UoECzgEACAAJ [Accessed 5/21/2021].

Pien, L.-C., Chen, D.-R., Chen, C.-J., Liang, K.-M. & Cheng, Y. 2015. Associations of neighborhood-level workplace violence with workers' mental distress problems: A multi-level analysis of Taiwanese employees. *Journal of Occupational Health*, 57, 555–564.

Razavi, S. & Turquet, L. 2016. Progress of the world's women 2015–2016: Transforming economies, realizing rights. *Global Social Policy*, 16, 86–93.

Scarselli, A., Corfiati, M., DI Marzio, D., Marinaccio, A. & Iavicoli, S. 2018. Gender differences in occupational exposure to carcinogens among Italian workers. *BMC Public Health,* 18, 413.

Smith, P. R. 2011. The pitfalls of home: Protecting the health and safety of paid domestics. *Canadian Journal of Women & the Law, Forthcoming, Washington University in St. Louis Legal Studies Research Paper*, 03.

Tollefsen, E., Langhammer, A., Romundstad, P., Bjermer, L., Johnsen, R. & Holmen, T. L. 2007. Female gender is associated with higher incidence and more stable respiratory symptoms during adolescence. *Respiratory Medicine*, 101, 896–902.

UN. 1995. *Beijing declaration and platform for action* [Online]. New York: United Nations. Available: www.un.org/womenwatch/daw/beijing/pdf/BDPfA%20E.pdf [Accessed 5/21/2021].

UN Women. 2020. *The shadow pandemic: Violence against women during COVID-19* [Online]. New York: UN Women. Available: www.unwomen.org/en/news/in-focus/in-focus-gender-equality-in-covid-19-response/violence-against-women-during-covid-19 [Accessed 5/21/2021].

UNDESA. 1997. *Mainstreaming the gender perspective into all policies and programmes in the United Nations System, report of the Secretary-General* [Online]. New York: United Nations. Available: www.un.org/esa/documents/ecosoc/docs/1997/e1997-66.htm [Accessed 5/21/2021].

UNGA. 1979. Convention on the elimination of all forms of discrimination against women.

UNGA. 2015. Sustainable development goals. *SDGs Transform Our World*, 2030.

Walsh, D. C. & Kelleher, S. E. 1987. The "corporate perspective" on the health of women at work. *Women and Work: An Annual Review.*

Weissman, A. M., Hartz, A. J., Hansen, M. D. & Johnson, S. R. 2004. The natural history of primary dysmenorrhoea: A longitudinal study. *BJOG: An International Journal of Obstetrics & Gynaecology*, 111, 345–352.

WHO. 1995. *Global strategy for occupational health for all: The way to health at work* [Online]. Geneva: World Health Organization. Available: www.who.int/occupational_health/globstrategy/en/ [Accessed 5/21/2021].

WHO. 2008. *Closing the gap in a generation: Health equity through action on the social determinants of health: Final report of the commission on social determinants of health* [Online]. Geneva: World Health Organization. [Accessed 5/21/2021].

WHO. 2011a. *Building healthy and equitable workplaces for women and men: A resource for employers and worker representatives* [Online]. Geneva: World Health Organization. Available: www.who.int/occupational_health/publications/healthy_workplace_9789241501729/en/ [Accessed 5/21/2021].

WHO. 2011b. *Gender mainstreaming in WHO: What is next?: Report of the midterm review of the WHO gender strategy 2011* [Online]. Geneva: World Health Organization. Available: https://apps.who.int/iris/bitstream/handle/10665/44709/9789241502337_eng.pdf?sequence=1&isAllowed=y [Accessed 5/21/2021].

WHO. 2013. *WHO global plan of action on workers health (2008–2017)* [Online]. Geneva: World Health Organization. Available: www.who.int/occupational_health/who_workers_health_web.pdf [Accessed 5/21/2021].

WHO. 2021. *Global health observatory* [Online]. Geneva: World Health Organization. Available: www.who.int/data/gho [Accessed 5/21/2021].

5

GENDER INEQUITY RELATED TO CLIMATE RISKS

Cultural vulnerabilities as obstacles to sustainability in rural communities of Mexico

Cloe Mirenda and Elena Lazos-Chavero

> Before, it rained a lot in June, but not anymore. Now it rains very late. Before, we planted June 3, and September 3 there was maize. Now we plant later. Now there are a lot of landslides and it affects the roots. Now there's not enough maize. The family suffers.
>
> *interview with Jacinta, Tatahuicapan, Veracruz, July, 2019*

> It's gotten really hot. It rains little. For example, now we're in the rainy season, but it hasn't rained. It has rained a bit but we've mostly felt a lot of heat. The watering holes have dried up. That's how we realized it had been very hot. Before, the women went to fish. Now there's nothing. The rivers are dry without fish, and us without food.
>
> *interview with Valentina, Tatahuicapan, Veracruz, July, 2019*

> Twenty years ago it rained more. This time of year, the streams were rising. The soil was very wet. Not like now, when it's dry. Now we can't even plant maize.
>
> *interview with Juana, Tatahuicapan, Veracruz, July, 2019*

> This year it hardly rained, like three, four times. That's why people hardly sow anymore. Because the plant dries up. And there are times when it rains a lot and the water milpa drowns.
>
> *interview with Margarita, Pérula, Jalisco, September, 2018*

5.1 Introduction

Much of the literature recognizes that climate threats and disasters have a differential impact on the population depending on the ecological, economic, and social

DOI: 10.4324/9781003174257-6

vulnerabilities that have accumulated in the region over time (Cannon 1994, 2000; Blaikie et al., 1996; Wisner et al., 2003; Cutter and Finch, 2008), including vulnerabilities based on women's and men's differential technical, economic, and social capabilities that predispose them to higher or lower levels of climatic risk (Aguilar et al., 2009; Soares and Murillo, 2013; Ímaz et al., 2014; Chavez-Rodriguez, 2016; Vázquez García, 2017). Some studies of climate disasters have focused on economic inequality, extreme poverty, development of social or mutual aid networks, differences in educational level, lack of democracy, and corruption. However, little research has addressed the role of cultural vulnerabilities associated with gender inequity[1] in the face of climate risks and disasters. Several authors have pointed to social, economic, and environmental dimensions of vulnerability, but we want to highlight the gendered cultural differences, considering cultural vulnerability as a set of relationships that have been reproduced historically in contexts of domination and violence (Wilches Chaux, 1993 Mirenda and Lazos, 2021). In the face of extreme climate phenomena, women's cultural vulnerability is evident in their lack of alternatives upon not having land or the same formal educational levels as men, which would provide them with equal access to labour markets (Mirenda and Lazos, 2021).

The aim of this chapter is to understand, through the analysis of the Sustainable Development Goals (SDGs) framework, how cultural gendered vulnerabilities influence the risk of climate threats and disasters in two communities on Mexico's Pacific west coast in the state of Jalisco, as well as in a Nahua community on the east coast in the state of Veracruz. We analyse how local gender inequity – understood here as differential access to land and territory, lack of access to decision-making institutions, and disparities in political power – magnifies women's social, economic, and political vulnerabilities to climate risks and disasters. In the context of contemporary rural Mexico, where the incidence and intensity of extreme climate phenomena (i.e. hurricanes, droughts, heat waves) have increased in recent years, gendered cultural vulnerabilities heighten their risk to such threats. It is important not only to exhibit these inequities but also to establish how historically these vulnerabilities have been constructed and expressed in different socioecological contexts.

Cultural vulnerabilities that provoke gender inequities in the face of climate disasters are historically reproduced in political, social, and economic spheres. We attempt to identify some of the consequences of women's lack of political participation and masculinization of power on a local level by taking into consideration SDG Target 5.5 ("Ensure women's full and effective participation and equal opportunities for leadership at all levels of decision-making in political, economic, and public life") and SDG Target 10.2 ("By 2030, empower and promote social, economic, and political inclusion of all, irrespective of age, sex, disability, race, ethnicity, origin, religion, or economic or other status"). With respect to SDG Target 5A ("Undertake reforms to give women equal rights to economic resources, as well as access to ownership and control over land and other forms of property, financial services, inheritance, and natural resources, in accordance with national laws"), we argue that while the legal framework provides for gender inclusiveness, this can be driven only from an essentialist perspective that dissociates women's recognition as subjects with rights, and besides, in practice, on a local level gender inequity in access to land persists due to inheritance traditions. This study, regarding gendered vulnerability in the face of climate risks and disasters, highlights the importance of building capacities (for example, related to political participation, organizing, and defence of human rights) and awareness-raising through education on gender equality as one pathway to sustainability, as proposed in SDG Target 13.3 ("Improve education, awareness-raising and human and institutional capacity on climate change mitigation, adaptation, impact reduction and early warning"). Nevertheless, it becomes critical to question the terms and the ways to foster the political participation and organization of women, in order not to follow the steps of women's integration to development of the 1980s. As Kabeer (1994) stated, "not to cure cancer with a band aid."

5.2 Methods

This study, which used qualitative methodology, consisted of several phases of fieldwork from January 2017 to November 2019. During the first phase of fieldwork (2017–2018), 40 semi-structured interviews were conducted with men and women in the rural communities of Arroyo Seco and Pérula, on Mexico's west coast in the state of Jalisco. Based on these interviews, a questionnaire was developed with the purpose of collecting data on men's and women's differential vulnerabilities related to climate risks and disasters, addressing the following topics: (1) household decision-making, (2) paid and unpaid work, (3) property, (4) political and social participation, (5) health, and (6) empowerment. During the second phase of fieldwork (2019), a questionnaire was applied to 40 people: 20 women and 20 men of these communities.[2] For the interviews and questionnaires, we selected adults (over the age of 18) who lived in the communities at the time of at least one of the hurricanes that had greatly affected the area in recent years: Hurricane Jova in 2011 and Hurricane Patricia in 2015. Additionally, 59 interviews (92% men, aged 33 to 83 and 8% women, aged 47 to 70) were carried out in the *ejido*[3] of Tatahuicapan de Juárez, on Mexico's east coast in the state of Veracruz, in July 2019. However, the majority of those interviewed in Tatahuicapan were accompanied by other men and/or women, and therefore we consider these to be group – rather than individual – interviews. The aim of the questionnaire was to comprehend men and women's experience of vulnerability to climate change, agricultural transformation, differential access to land according to gender and generation, and youth migration.

The community Arroyo Seco has fewer than 400 inhabitants (INEGI, 2010) and belongs to the *ejido* Ley Federal de Reforma Agraria. Pérula has approximately 800 inhabitants (INEGI, 2010) and belongs to the *ejido* La Fortuna. Residents of these communities are mestizo. The men principally carry out commercial crop agriculture, cattle raising, fishing, and construction, and women carry out domestic work in their homes and work in tourism and – to a lesser extent – crop agriculture. Since the 1990s, many members of these communities have migrated to the United States. We selected these two coastal communities as they have been strongly affected by two hurricanes within a short period of time: Jova (category 2) and Patricia (category 4), and more recently by less intense climatic events such as Cyclone Narda and Hurricane Lorena (category 1) in 2019. In recent decades, the incidence and magnitude of extreme meteorological events has increased in dry tropical forest regions throughout Mexico's Pacific Coast (Jáuregui, 2003; Álvarez-Yépiz and Martínez-Yrízar, 2015; Maass et al., 2018), and it is likely that due to global warming the intensity and frequency of these events will continue to increase (IPCC, 2014).

In the Sierra of Santa Marta of southern Veracruz, the study community Tatahuicapan de Juárez[4] belongs to the municipality of the same name. While the original inhabitants were of Nahua origin, many mestizos migrated to Tatahuicapan from central Veracruz in the late 1960s as a result of the colonization programme "March toward the sea." As a result of influence of the mestizos, who developed cattle raising as their dominant economic model, the Nahua residents gradually transformed their *milpas* (diversified polyculture of maize, beans, squash, and other vegetables) into low-productivity pastures. This transformation, along with extreme climatic phenomena, have exacerbated dietary poverty since the 1990s in the municipality of Tatahuicapan. While in 2010, 40.9% of the population lived in a condition of food poverty, in 2015 this had increased to 51.2% (CONEVAL, 2019). Several statements by Tatahuicapan women express the relationship between reduction in rainfall and reduction in the volume of rivers, which in turn has provoked reduction in fish and shrimp that had been important to their diet (see quotes at the beginning).

5.3 Theoretical perspectives

The contextual vulnerability approach defines climate vulnerability as a multidimensional historical process consisting of "the characteristics of a person or group and their economic, social and political situation influencing their capacity to anticipate, cope with, resist and recover from the impact of a natural hazard" (Blaikie et al., 1996; Wisner et al., 2003: 4, 11). This interpretation forces us to reflect on the importance of using a historical perspective to analyse processes which lead to vulnerability (Copans, 1983; García Acosta, 2008; Ribot, 2011, Taylor, 2015). Cannon (1994:14) points out the existence of "sets of unequal access to opportunities and unequal exposures to risks which are a consequence of the socioeconomic system." Such processes have developed over time through social and socio-environmental relations, producing disasters whose effects are not always directly proportional to the frequency and magnitude of natural phenomena. Taylor (2015) expresses the need for a historical methodology capable of tracing how contemporary experiences of climate change interfere with social, economic, political, environmental, and cultural processes already underway. From this perspective, rather than attempting to evaluate which people are vulnerable to climate risks by simply considering vulnerability and adaptive capacity to be intrinsic properties of individuals, we must also explore production and reproduction of vulnerabilities provoked by climate risks and disasters as the expression of complex, dynamic socio-ecological relations among social classes, genders, generations, ethnical, and other social groups in different socioecological contexts (Taylor, 2015; Fosado, 2019). According to a gender perspective, different forms of socially constructed vulnerabilities are derived from the social roles that men and women assume – or which are imposed upon them, as well as patterns of gender inequity and discrimination (Neumayer and Plümper, 2007). Depending on the context, such vulnerabilities and inequity are manifested in women having little access to healthcare and food; lacking mobility; being granted less decision-making capacity than men; experiencing inequity in legal, market, and political institutions; and generally being granted a lower level of human rights than men. These inequities merge to form a complex web of gendered vulnerabilities to climate threats and disasters in the decision-making processes before, during, and after the event (Juran and Trivedi, 2015: 602).

Following a feminist political ecology perspective, we focused on three key causes of differential vulnerabilities: (1) Gendered land and territorial access; (2) Gendered division of labour within the household, including gender differences in power with respect to use, control, and ownership of resources which results in lack of access to decision-making; and (3) Gendered access to community and regional politics.

In the light of these theoretical considerations, we structure our analysis by addressing the historical processes that shape livelihood opportunities and distribution of power in family and community decision-making. Subsequently, we analyse the implications of these processes on accumulation of types of vulnerability to climate risks and disasters differentiated by gender.

5.4 Gender mainstreaming in risk management, climate change, and sustainability

International policy has advanced significantly with respect to considering gender in documents on sustainable development, risk management, and climate change, although many of them continue to have a simplistic gender perspective, describing only the distribution of activities, opportunities, and behaviours. The intersectional approach allows us to understand the articulation between gender, social class, ethnic groups, and generations in different political and socioecological contexts, which will enrich the analysis of gender and climate vulnerabilities. Since the 1992 Rio Declaration, a multitude of governments, NGOs, social movements, and others have called for recognition of gender in all development and environmental policies and programmes, as reflected in the current version of the UN SDGs, which include gender equality and empowerment of women and girls (SDG 5). UN risk management frameworks seek to take into account the multiple socioeconomic, political, and environmental causes of disasters through an approach that recognizes that disaster prevention strategies must be based on reducing vulnerability by promoting sustainable equitable human development (Lucatello, 2008: 68). The preventive approach to risk reduction began to be considered during the International Decade for Natural Disaster Reduction 1990–2000 and was reinforced during the World Conference on Disasters in Kobe (2005) with the presentation of the Hyogo Framework for Action. This document calls for integrating a gender perspective in all phases of the disaster cycle: risk assessment, early warning, preparedness, response, and recovery (Aguilar et al., 2009: 55). The Sendai Framework for Disaster Risk Reduction 2015–2030 also emphasizes gender demands, specifically calling for greater leadership of women in disaster risk reduction, as well as generation of data on disasters disaggregated by sex to build evidence of disproportionate impacts on women and girls with the purpose of developing strategies to minimize these impacts.

In recent decades, a preventive approach to disaster reduction based on the concept of comprehensive risk management has gradually been integrated into Mexico's legal framework. Mexico's General Civil Protection Law (Gobierno de la República, LGPC, 2012) defines comprehensive risk management as the "set of actions aimed at identification, analysis, evaluation, control, and reduction of risks, considering their multifactorial origin in a permanent process of construction" and is related to sustainable development to combat the structural causes of disasters (Art. 4). However, despite this shift towards preventive multifactorial risk reduction, gender does not appear even once in this law, and in Mexico's 2014–2018 National Civil Protection Plan, a gender perspective is mentioned only three times: in the guidelines of action, creation of legal systems which are complementary to the LGPC, and promotion of research on comprehensive risk management. One significant preventive action – creation of the Disaster Prevention Fund (FOPREDEN according to its Spanish acronym) – provides funds to public agencies to implement risk reduction actions. Other prevention mechanisms include Agroaseguradora Mexicana, a government agency that provides insurance for the rural populations, the Comprehensive Program for Sustainable Agriculture and Productive Reconversion in Recurrent Accident Areas, and the Fund to Serve the Rural Population Affected by Extreme Weather Conditions, which includes crop insurance (Oropeza Orozco, 2008: 135).

While Mexican law and policy has begun to focus on disaster prevention, actions carried out with this purpose appear to contradict this principle. Disaster management schemes continue to be reactive rather than preventative in nature, based on military "aid" and temporary measures, regarding disasters as a national security problem (Rodríguez Velázquez, 2011: 130). According to this approach, when a disaster occurs, a state of exception is declared, thereby empowering the State not only to implement emergency actions that do not reduce vulnerability (Rodríguez Velázquez, 2011), but also to intervene in an authoritarian manner, even involving the military. As a result, a context of anomie is created in which the State justifies violence and establishes conditions for "silenced production of social catastrophes"

(Marchezini, 2014). The state of exception – a mechanism common throughout Latin America – allows disaster to be managed as an emergency, declaring the validity of controlling the social order through actions such as suspension of legal regulations, which allows for violation of human rights (Marchezini, 2014). Additionally, this model fosters the erroneous idea that the population affected by a disaster is completely defenceless. Finally, by leaving the task of attention to disaster as a matter of national security for which the federal government is fully responsible, responses to disaster are limited to piecemeal reactive, rather than preventive, measures (Rodríguez Velázquez, 2011).

Mexico's principal programme that uses this approach has been the DN-III-E Plan for Assistance to the Population in Case of Disaster, created in 1966 by the Secretary of National Defence following floods in Tampico, Tamaulipas (Rodríguez Velázquez, 2011). This plan has generated public acceptance that those who are capable of acting in the event of disasters are the armed forces and led the federal government to apply an ideology of authority and control to avoid public actions that they consider might threaten the institutional order (Rodríguez Velázquez, 2008: 41). As has been analysed from a feminist constructivist approach, such an ideology is framed in dominant gendered discourses (MacGregor, 2010: 229). Disaster risk management policies – and more recently climate change policies – tend to be framed in the typically masculinized process of securitization and militarization of the environment. Climate risks – seen as threats to national and global security – require technical, diplomatic, and military solutions that appear to be in total accordance with stereotypes of hegemonic masculinity (MacGregor, 2010).

Nonetheless, the issue of gender is gradually beginning to be recognized within climate change policy, as analysed by several authors (Aboud, 2011; Skinner, 2011; Vázquez García, 2015; Aguilar et al., 2017; Fosado, 2019). Initially absent in the documents of the United Nations Framework Convention on Climate Change (UNFCCC), including the Kyoto Protocol, during the past decade, the issue of gender has begun to be integrated into climate change policy. In 2007, during COP 13 in Bali, a meeting of government ministers called for action regarding gender equality within the framework of human rights and international agreements (Red de Género y Medio Ambiente, 2010). On this occasion, the International Union for Conservation of Nature (IUCN), the United Nations Development Program (UNDP), the United Nations Environment Program (UNEP), and the Women's Environment & Development Organization (WEDO) founded the Global Gender and Climate Change Alliance (GGCA). Even though its enormous limitations, in 2009, this alliance, together with the Government of Finland, created the Fund for Women Delegates to increase women's participation in national delegations to international climate change negotiations and close gender gaps through financing capacity building, and development of networks, outreach, and advocacy (Aguilar et al., 2017: 59). In 2011, the UN Commission on the Legal and Social Status of Women adopted its first resolution addressing the connection between gender equality and climate change, and in 2014, it approved another resolution on gender equality and women's empowerment in the context of disasters (Aguilar et al., 2017).

In Mexico, government mechanisms for climate change reduction formally reflect these advances. The General Law on Climate Change establishes that state programmes must promote gender equity and representation of those populations most vulnerable to climate change (LGCC, Art. 17). Based on the principles of this law, the National Climate Change Strategy (ENCC, according to its Spanish acronym) includes the following guideline of action: "In designing all climate change policies, consider gender, ethnicity, disability, inequality, health status, and inequity in access to public services, and involve different sectors of society in their implementation." The ENCC proposes to include a gender approach in strategies for reducing social vulnerabilities (A1.11). The Special Program on Climate Change (PECC 2014–2018) also reflects the need to address vulnerability to – and effects of – climate change differentiated by gender through various objectives and strategies of the National Program for Equal Opportunities and Non-Discrimination Against Women (PROIGUALDAD 2013–2018). This programme also indicates that the Climate Change Fund and other climate related funds must prioritize gender equity, transparency, and efficiency (Vázquez García et al., 2015). Although its essentialist perspective and its limitations, there has been a slow recognition of gendered vulnerabilities as the lack of recognition of women as subjects of law and rights.

5.5 Historical roots of cultural vulnerability from a gender perspective in Jalisco and Veracruz

In Jalisco, the communities Arroyo Seco and Pérula were founded in the 1960s as a result of Mexico's Agrarian Reform, which attempted to comply with the demands of the Mexican Revolution to provide landless peasants with land under the motto "the land belongs to those who work it." Prior to the Agrarian Reform, the Nahua people of Tatahuicapan, Veracruz held communal land tenancy. Unequal land distribution led to serious conflicts among community members, resulting in government redistribution of land in the 1970s.

The original Agrarian Reform was characterized by a sexist bias; according to the Agrarian Law of 1927 and 1934, all adult men could be granted land regardless of their marital status and number of children, while women could receive land only if they were widows or single mothers (Vázquez, 2016: 175). For this reason, the first founders of the *ejidos* were men, while women arrived to the communities usually a year later, as wives or daughters of the original founding peasants, but were never granted land rights. Such gender discrimination by Agrarian Law is characteristic of the broader legal system established by the Mexican Constitution. In 1937, citizenship was denied to women, thereby evidencing resistance to recognizing equal rights for women and men (Mondragón, 2012: 203). Not until 1947 were women granted the right to vote and to be voted for – although only in municipal elections, and only in 1953 was full citizenship recognized (Mondragón, 2012: 203).

In 1992, *ejido* regulations were modified and the Program for the Certification of Ejido Rights and Titling of Urban Plots (PROCEDE according to its Spanish acronym) was implemented, legalizing sale of *ejido* land. This reform did not establish gender distinctions for land tenancy. As a result, the number of women owning land increased (Warman, 2003). However, it has also been argued that these changes did not effectively benefit women, given that they have limited purchasing power. Furthermore, due to the fact that PROCEDE also allowed people outside the family to inherit land from *ejido* members, women's legal opportunities to become heirs were reduced (Vázquez, 2016: 188). In the study of communities of Jalisco and Veracruz, women landholders are still a small minority and acquire land only as widows of *ejido* members. However, this generally occurs only when women are elderly and can no longer work the land, resulting in two possible outcomes. First, the female *ejido* member is in charge of preserving family patrimony that will be transferred in the future to one or more adult male children. That is, she is just a link in the male transfer of land (Vázquez, 2016: 181). The second possibility is that if no sons or daughters are interested in working the land, the woman ends up selling the land. In Arroyo Seco, only 1 of 48 *ejido* members is female, and in La Fortuna, of approximately 70 *ejido* members, fewer than 10 are women. In Tatahuicapan, of 767 *ejido* members, approximately 60 are women. Thus, in the study area, women make up fewer than 10% of all *ejido* members. Additionally, in Jalisco, the idea persists that working in the field is a male activity. In Veracruz, women formerly worked in the *milpa*, but this practice ended with the transition to cattle raising. Men control decision-making regarding planning, raising, and sale of cattle. As a result, masculinization of land access and ownership has reduced women's ability to adapt their livelihoods to a changing climate (IPCC, 2007; Vázquez et al., 2015).

Furthermore, the 1992 Agrarian Reform law has increased the vulnerability of the *ejido* in general by modifying the relationships among *ejido* members as well as the relationship between them and the land. Due to its impact, some authors consider this reform to have marked the end of collective land tenancy in Mexico (Pérez Castañeda et al., 2015). As they now had an individual land title, some *ejido* members began to sell their plots without authorization by the community assembly. In other areas of Mexico, it has been documented that this process increased the vulnerability of *ejido* members to the extent that it affected their capacity for collective agency which had previously allowed them to adapt to changes, rather generating a subjectivity of addressing risks individually (Richard, 2008). In the *ejidos* La Fortuna, Ley Federal de Reforma Agraria, and Tatahuicapan, this process of individualization and fragmentation of land has been observed to have resulted in a progressive decrease in participation in community assemblies and collective labour.

While gradual loss of power of the community assembly was exacerbated by PROCEDE, the origin of its weakening lies in the process of land distribution, which was not always equitable in the study areas, as in other parts of Mexico. *Ejido* members with larger landholdings co-opted leadership positions in the *ejido*, leading to a general loss of interest in participating in assemblies, as the last word in decision-making was in the hands of a few. Currently, in Jalisco, a few peasants attend the assembly, to the point that they are held every two months instead of the traditional monthly meetings. These assemblies, which were once the collective heart of the *ejido*, are losing their function, in turn increasing division and conflicts among *ejido* members. In Arroyo Seco and Pérula, it was commented that some *ejido* members do not follow the procedure of requesting permission from the assembly to sell their plots, and thus the *ejido* does not receive the established percentage on its sale.

In Tatahuicapan, the assembly still has political force, although this has been reduced by individualization of land. Nevertheless, due to the presence of the Yuribia Dam on their land since the 1990s, the *ejido* assembly of Tatahuicapan still makes some decisions collectively regarding water management and government agricultural programmes.

With respect to agriculture after the 1960s, when subsistence production predominated, peasants in Jalisco began to produce commercial crops such as coconut, watermelon, lemon, chili, papaya, and mango. However, studies since the 1940s have indicated that due to erratic rainfall in the region, large-scale agricultural and livestock production is inviable (Castillo et al., 2014: 384). Agrarian policy encouraged such production during the 1970s and 1980s, when policy discourse considered forests to be useless land, and many dry tropical forest areas were thereby converted into cropland and pastures.

Such agricultural policies led not only to ecological damage to large portions of forest, but also to unprofitable agriculture which provoked migration of many people (Castillo et al., 2014: 383). Having later abandoned crop production due to poor harvests, women who had moved to these communities due to high demand for female labour now have fewer job opportunities. As the following testimony shows, the effects of failed agricultural policies on the Coast of Jalisco are still felt today:

> My mother-in-law told me there was a lot of work for women planting and harvesting chili peppers, and I came with my husband. We lived in Tecuán [...] Now there's no more work in the fields. They rarely invite you to work. They stopped planting chili, they stopped planting tomatoes. [The harvest] started to decrease, to be attacked by pests, and now there's only papaya, watermelon. It's no longer the same job because now they hire a lot of men to plant, to harvest, and before it was more even.
>
> *woman, 53 years old, Arroyo Seco, November, 2017*

Decreases in productivity and pests have led to reduction in employment opportunities, especially for women, affecting rural livelihoods in a gender-differentiated manner. On the southern coast of Jalisco, government policy currently promotes tourism, which has various negative impacts on local residents, such as increased privatization of land, dispossession of natural resources, and abuse of power, as well as a loss of peasant identity (Gerritsen and Lazos, 2015: 3). However, promotion of tourism also has positive impacts from a gender equity perspective, as it is increasing job opportunities for women historically excluded from the region's principal economic activities. Although their work continues to be short term and poorly remunerated, and they lack social security, women consider tourism to be a labour opportunity. In Tatahuicapan, women have almost completely lost their role in their family's agricultural production. Due to the distance of agricultural plots from their homes and transformation of crop agriculture into cattle raising, many women have been largely displaced from family agriculture. Their decision-making power is restricted to certain stages during the harvest, and they do not decide on sale of the harvest. If they wish to retain harvested products for family consumption, the final decision is made by the men. Very few women have their own livestock. In Tatahuicapan, no job offers exist for women. Those women who had access to formal education and political support have been able to obtain teaching positions. However, the large majority of women carry out domestic labour with no remuneration or have no other job alternative but to migrate to northern Mexico. Furthermore, this precarious situation is culturally normalized. "That's the custom here – the women in the home," commented one livestock raiser, father of four daughters from the municipality of Tatahuicapan. Very few fathers seek to provide alternatives for their daughters. One of the few who do so is another livestock raiser from Tatahuicapan, who states,

> As I only have two daughters, I taught them to round up cattle, to milk, to fix the paddock. They also know how to make cheese and go sell it. I just need to teach them to drive so they can go to Coatza[coalcos, Veracruz] to sell the cheese.

In the coastal communities of Jalisco, as in the Nahua village of Tatahuicapan in Veracruz, we found that extreme climate phenomena, low agricultural prices, and migration provoked agricultural changes and loss of dietary diversity.[5] The women of these communities have lost access to previously female-dominated areas of food procurement due to deforestation; disappearance of much riparian fauna due to environmental changes and water pollution; and replacement of *milpas* with commercial agricultural systems or pastures. This has made them almost completely dependent on male-dominated agricultural systems (livestock raising and commercial crop agriculture) and purchase of food (with money sent by migrants or from federal welfare programmes). The resulting subordination and lack of access to land provoke gendered vulnerabilities. Furthermore, women culturally depend on decision-making by the male head of family (husband, son, grandson, father, or grandfather), which increases their vulnerabilities.

5.6 Control over land and property

With respect to SDG Target 5A (Undertake reforms to give women equal rights to economic resources, access to ownership and control over land and other forms of property, financial services, inheritance and natural resources, in accordance with national laws), we argue that gender inequity persists with respect to their living conditions. The power to make decisions regarding family property and assets is a significant factor in post-disaster recovery. As the present study indicates, male property owners make relevant decisions individually, very rarely consulting their partner regarding land improvements, investments, and purchase or sales. In the communities of Jalisco, 35% of the women but none of the men surveyed live in houses owned by their partners, and therefore women are subordinated to the partner's decisions to a greater extent than the men. Among those surveyed, only a minority own the home in which they live: 25% of women and 30% of men. In these communities, a gender difference exists in the relationship between home ownership and marital status: women owners are mainly widows, while men are property owners regardless of their marital status. The rest live in the homes of relatives or employers (30% of women and 50% of men) or in homes without deeds (10% of women and 20% of men).

In Tatahuicapan, the majority of women (over 70%) live in houses owned by their husbands or by their husband's parents or other relatives. As a Nahua community, patrilocality is a well-established kinship and residence system. However, due to recent migration, women are gradually becoming homeowners.

In the communities of Jalisco, people who work as caretakers or hotel employees and live on their employers' property may find themselves in a vulnerable situation as they depend on the will of the owners to repair damages caused, for example, by hurricanes. One woman who lives in a cheap hotel in Pérula where she works as a housekeeper reports the difficulties she faced after Hurricane Jova. She was not provided with sheet metal roofing donated by the government because they were exclusively destined for housing, not for tourism. As the owner was reluctant to make necessary repairs, despite considering it to be unfair that she invest her savings in property that does not belong to her, she was forced to use her own money to replace roofing destroyed by the hurricane.

Land ownership is another factor that intervenes in post-disaster recovery. In the study communities of Jalisco, 20% of women and 35% of men surveyed own land. However, gender differences exist in strategies for coping with hurricane damage: most of the men sold a portion of their properties, while the women preferred to keep their property and resolve economic needs through savings or loans.

5.7 Male domination of political participation

SDG Target 10.2. proposes to promote economic and political inclusion of all, irrespective of age, sex, disability, race, ethnicity, origin, religion, or economic or other status. Nevertheless, the majority of Mexico's rural communities are characterized by a lack of political participation with gender equity and masculinization of power. Gender-differentiated vulnerabilities are related to women's lesser political capital due to their isolation and the fact that they have fewer opportunities to organize (Vázquez García et al., 2015).

In Pérula and Arroyo Seco, in Jalisco, women participate less than men in community and political organization, and people of both sexes were generally found to be against the idea of assigning political positions to women. Furthermore, even though women appear to have gained political participation, male patterns in the exercise of power – which have historically generated profound vulnerabilities for women in the region – continue to predominate (Mirenda and Lazos, 2021). For example, lack of acceptance of women in politics is ostensibly justified by their lack of formal education and experience; because of the image of women's proper role in their homes and communities, which does not include holding political positions; and because of acceptance of a macho culture in politics that excludes women's participation.

In Tatahuicapan, there has never been a female *ejido* elected community assembly leader (*comisaria* in Spanish). The simple fact of asking if during the *ejido*'s history a woman has held the position of commissariat (*comisaria*) results in mocking and negative comments regarding women's political role. Due to the low number of female *ejido* members given that typically they are only *ejido* members while they regularize the deed for the sons, there is no women's organization to demand their rights. Women denigrate themselves due to lack of formal education, experience, and knowledge required to carry out political functions. Only three or four female *ejido* members have intervened temporarily in political roles. As one female *ejido* member of Tatahuicapan asked, "[...] how are we going to learn, if they don't let us?"

5.8 Climate action

This is related to SDG Target 13.3 (Improve education, awareness-raising, and human and institutional capacity on climate change mitigation, adaptation, impact reduction and early warning). Nevertheless, education and generation of institutional capacities to confront climate change are not priorities of federal or state governments, which do not even pose this objective as a long-term goal.

Despite international advances in considering gender equality in the context of management of climate risks, actions by the Mexican government are highly inconsistent in the face of disasters. Those rural communities which tend to face greater climate risks depend on the political will of each state to declare a "state of disaster" and place into effect the DN-III-E Plan for Assistance to the Population in Case of Disaster. Despite the fact that preventative plans are mandated by law, no such plan exists for the study areas. Nor does any collaboration take place among state, municipal, and local governments. On the contrary, in the case of a disaster, political clashes occur if the authorities of different geographic levels belong to different political parties. Such lack of prevention further intensifies existing inequalities among rural and urban communities, social classes, generations, and genders. "Assistance" is left to the good will of each state (particularly through purchase of votes by providing subsidies to offset damage) and the military.

5.9 Discussion and conclusions

In Mexico, the big challenge for sustainability and for achieving SDG 5 (Gender equality; in particular Target 5.5. and Target 5A), SDG 10 (Reduced Inequalities), and SDG 13 (Climate Action) is first to recognize that these SDGs must be implemented into the legal framework and practices through public debate and policies in the recognition of women as subjects of rights. When sustainability and gender equity have been addressed, it has been separately, as if they were not related. The underlying drivers of unsustainability, social injustice, and gender inequity are closely related (Leach et al., 2016). They are produced by political-economic and cultural dynamics that support a patriarchal neoliberal system based on market-led growth and social inequalities. As we have analysed, gender inequality is being considered in federal laws and policies regarding sustainability, but typically they are only expressed as idealized or simple concepts, without considering the intersectionality with other factors (social classes, generational inequities, ethnic differences). Policy debate is lacking to be able to implement laws to reduce gender inequality locally.

Many forms of inequity are continuously reproduced by cultural and social norms and institutions dominated and controlled by men, including gender inequity, exploitation of women's labour, provision of unpaid domestic care, lack of access to land and food sources, lack of access to decision-making regarding their agricultural plots, the sale and destination of agricultural products, lack of access to formal and informal community institutions, and lack of access to political participation. In the study communities, these inequalities are continually reproduced and exacerbated by economic and social inequalities, extreme poverty, and hunger, thereby increasing cultural, economic, and social vulnerability of women to climate risks and threats. To understand how women are trapped in the male decision-making processes during and after a disaster because of their economic or educational or social contexts allows us to consider the differentiated consequences of climate risks and disasters among men/women/children/elder people/ethnic groups. Cultural vulnerability is reproduced upon assigning specific roles and functions to each gender. Sexual division of labour is central to reproduction of cultural vulnerability, assigning women to the domestic sphere or to lower paid labour than that of men. As a result, women lack secure access to housing where they can make decisions before, during, and after a climate disaster. Furthermore, they are unable to make decisions regarding designation of crops for sale as they do not have secure access to land. Even, women cannot decide to sell a cow in order to acquire her medications; they depend on their husband's or sons' decisions. Finally, given cultural gender roles, it is difficult for them to develop organizational capacities to confront climate risks (e. g. droughts that condition crop production and therefore the access to their annual consumption) and disasters (of all kinds, from climatic to social or as nowadays health ones) as they don't have access to the information nor the social relationships with technicians of different development or agrarian institutions that could provide some help.

We do not wish to provide an image of women solely as victims of environmental degradation, or – in this case – of negative impacts of climate change, as occurred in the WID (Women in Development) discussions of the 1980s. Debate that includes analysis from a feminist political economy perspective – especially regarding access to land, and political participation (Benería and Sen, 1981; Young et al., 1984) – has helped to reformulate development policies that previously produced gender inequity. As a result, models of gender justice now exist which incorporate access to land, water, forests, and jobs based on rights and capacities with gender equity (Rai and Waylen, 2013; Agarwal, 1994; Braidotti et al., 1994; Rocheleau et al., 1996). Analyses from a feminist political ecology perspective allow for taking into account intersections among ecological and climatic problems, as well as gendered power relations, from the household to the global level (Butler, 1994; Elmhirst, 2011), thereby allowing for new socio-environmental relationships characterized by gender equity.

In Mexico, movements for feminist struggle exist in rural/urban communities (e.g. National Encounters of Indigenous Women), providing lessons and narratives of change whereby women revindicate their rights in all spheres (Lazos and Mirenda, 2021). Future challenges include identifying such mechanisms for building synergies between development of women's social, political, and cultural capacities on one hand, and on the other, awareness-raising to be prepared to act in the face of climate threats and disasters with a gender perspective. In order to reduce gender inequality (SDG 5) and take urgent action against climate risks and threats (SDG 13), we must assure mechanisms for effective participation of rural men and women with equal opportunities to develop leadership in decision-making in the political and public life of their communities.

Acknowledgements

We thank the Nahua population of Tatahuicapan in southern Veracruz and the communities of Pérula and Arroyo Seco on the southern coast of Jalisco for sharing their thoughts and feelings regarding their experiences in facing climate risks and disasters. Thanks to financing by DGAPA of UNAM, we were able to carry out fieldwork under the guidelines of the project titled "Threats and vulnerabilities of Mexican agriculture: loss of agrobiodiversity and control of seeds, youth migration, and climate change" (PAPIIT IN304519) coordinated by Elena Lazos Chavero. We thank the project members for their collaboration in fieldwork: Marcela Jiménez, Palestina Luz Llamas Guzmán, Jazmín Solís Carpio, Tania Flores Gutiérrez, and Esteban Ramírez Cruz. We thank the biology research center, Estación de Biología de Chamela, of UNAM for hosting us during fieldwork. We would also like to thank Ann Greenberg for support with translation.

Notes

1 Prior to the 1995 IV World Conference on Women in Peking, the use of the terms equity and equality was highly debated. While conference participants decided to use equality, many Latin American feminists disagreed with this term as it has been understood to mean "being equal to men." Rather, they revindicated the term "equity," which emphasizes equal opportunities, or treatment for each gender according to their needs. Despite the fact that many policies have replaced the term equality with equity, little has been achieved in terms of actually achieving equity. As equality is a human right and therefore a legal obligation of nation states, many governments propose to recover the term equality, accompanied by "non-discrimination." Therefore, most documents by governments and multilateral organizations (including the SDGs) speak of equality. Even though, equality is a political principal of recognition and rights, and equity is one strategy of political policies to achieve it, the authors choose to maintain the term equity to emphasize struggle for equal opportunities for men, women, and people of other genders.

2 Ten percent of the adult population, based on data by the Institute of Statistical and Geographical Information of Jalisco (IIEG) and the Population and Housing Census of Mexico's National Institute for Statistics and Geography (INEGI; 2010). In some cases, the semi-structured interviews and the questionnaires were applied to the same elder persons, as they could provide more the historical backgrounds and their life experiences with disasters and climate change.

3 The *ejido* is a type of landholding particular to Mexico that originally consisted of communal landholding and was governed by principles protecting peasants' rights, assuring that land was inalienable, imprescriptible, and non-transferable (Ruiz Alarcón, 2015: 188; Pérez Castañeda et al., 2015). As a result of the 1992 Agrarian Reform, linked to NAFTA, these principles were largely abandoned.

4 The municipal seat, Tatahuicapan de Juárez, has 15,614 inhabitants, of whom 52% are women and 48% men (INEGI, 2016). The *ejido* has 11,234 hectares, with 466 original *ejido* members, but due to purchase of land as a result of the 1992 Agrarian Reform, currently there are 767 members (Registro Agrario Nacional, 2020), whose plot size ranges from 2 to 20 hectares (fieldwork 2019).

5 In a 1998 study carried out in Tatahuicapan, of 45 women interviewed, 75% of those over the age of 60 accessed the rainforest in search of food when they were younger than 30, but only 22% of them continued to do so. Similarly, while 44% of women aged 30–60 interviewed did so in the 1980s, by 1998, only 10% did so. In 1998, women aged 30 and younger did not go to the forest to gather food (Godínez and Lazos, 2003: 634). Women's food provisioning activities in the 1980s were gathering fruit and edible flowers in the jungle and fallows as well as a variety of mushroom species in oak forests; fishing and capturing snails, shrimp, and crabs in rivers and streams; and capturing edible insects (cicadas and occasionally grasshoppers); at that time, the majority of women over 60 (88%) and those aged 30–60 (75%) fished regularly in order to contribute to their family's diet (Lazos and Godínez, 2004: 634–635).

References

Aboud, G., 2011, *Gender and Climate Change Cutting Edge Pack*, Supporting Resources Collection. Retrieved from Sussex.

Agarwal, B., 1994, *A Field of One's Own: Gender and Land Rights in South Asia*. Cambridge: Cambridge University Press.

Aguilar, L., 2009, *Manual de capacitación en Género y Cambio Climático*. UICN, PNUD, GGCA.
Aguilar, L., Granat, M., and Owren, C. (eds.), 2017, *Las raíces del futuro: situación actual y progreso en género y cambio climático*, UNAM.
Alarcón Ruiz, A., 2015, "La organización ejidal en el desarrollo rural de México," *Estudios Agrarios*, 59, 181–204.
Álvarez-Yépiz, J. C. and Martínez-Yrízar, A., 2015, "Huracanes, sequías y heladas: eventos climáticos extremos en México," http://web.ecologia.unam.mx/oikos3.0/index.php/articulos/18-arch...climaticos-extremos-en-ecosistemas?tmpl=component&print=1&page=
Arias, P., 2009, *Del arraigo a la diáspora: dilemas de la familia rural*, México city, Universidad de Guadalajara, Centro Universitario de Ciencias Sociales y Humanidades, Miguel Ángel Porrua, H. Cámara de Diputados, LX Legislatura.
Benería, L., and Sen, G., 1981, "Accumulation, reproduction, and 'women's role in economic development', Boserup revisited," *Signs*, 7(2), 279–298.
Blaikie, P., Cannon, T., David, I. and Wisner, B., 1996, *Vulnerabilidad. El entorno social, político y económico de los desastres*, Red de Estudios Sociales en Prevención de Desastres en América Latina.
Braidotti, R., Charkiewicz, E., Häusler, S., and Wieringa, S., 1994, *Women, the Environment and Sustainable Development: Towards a Theoretical Synthesis*, London: Zed Books.
Butler, J., 1994, "Gender as performance: An interview with Judith Butler," *Radical Philosophy*, 67, 32–39.
Cannon, T., 1994, "Vulnerability analysis and the explanation of "natural" disasters." In A. Varley (ed.), *Disasters, Development and Environment*, Chichester: Wiley, 13–30.
Cannon, T., 2000, Vulnerability analysis and disasters. In D.J. Parker (ed.), *Floods*, London: Routledge, 45–55.
Castillo, A., Quesada M., Rodriguez, F., Anaya F., Galicia C., Monge F., Barbosa R., Zhouri A., Calvo-Alvarado J., and Sanchez-Azofeifa G. A., 2014, "Tropical Dry Forests in Latin America: Analyzing the History of Land Use and Present Socio-Ecological Struggles," in A. Sanchez-Afofeifa, J. Powers, M. Quesada (eds.), *Tropical Dry Forests in the Americas. Ecology, Conservation, and Management*, London, New York: CRC Press, Taylor & Francis Group, 357–394.
Chavez-Rodriguez, L., 2016, "La importancia de la interseccionalidad en la vulnerabilidad social ante eventos hidrometeorológicos extremos en Yucatán, México," en: Vázquez García, V., Velázquez Gutiérrez, M., Sosa Capistrán, D. M. y De Luca Zuria, A. (eds.) *Transformaciones ambientales e igualdad de género en América Latina: temas emergentes, estrategias y acciones*. Cuernavaca, Morelos: CRIM-UNAM, 19–42.
CONEVAL – Consejo Nacional de Evaluación de la Política de Desarrollo Social (2019), *Carencia por acceso a la alimentación por entidad federativa, 2018*, http://sistemas.coneval.org.mx/InfoPobreza/Pages/wfrMapaPobreza?pTipoPobreza=1&pTipoIndicador=1
Copans, J., 1983, "The Sahelian drought: social sciences and the political economy of underdevelopment," in K. Hewitt (ed.), *Interpretations of Calamity*, Boston, London: Sidney Allen & Unwin, 83–98.
Cutter, S.L., and Finch, C., 2008, Temporal and spatial changes in social vulnerability to natural hazards. *Proceedings of the National Academy of Sciences*, 105(7), 2301–2306.
Elmhirst, R., 2011, "Introducing new feminist political ecologies," *Geoforum*, 42(2), 129–132.
Fosado, E. J. (2019). "La lógica de género en la configuración del discurso climático: Un análisis del campo institucional (1994–2015)." *GénEros*, 26(26), 7–39.
García Acosta, V., 2008, "Riesgo y desastres ¿climáticos o sociales?, en la historia de la Cuenca de México," in A. Lammel, M. Goloubinoff, and E. Kats (eds.), *Aires y lluvias. Antropología*

del clima en México, Mexico and Paris, Centro de Investigaciones y Estudios Superiores en Antropología Social, Centro de Estudios Mexicanos y Centroamericanos, Institut de Recherche pour le Développments, 547–566.

Gerritsen, P., and Lazos-Chavero, E., 2015, "Transformaciones socio-ambientales y potencial agroecológico frente al cambio climático en la región Costa Sur de Jalisco, Occidente de México," *Memorias del V Congreso Latinoamericano de Agroecología*, La Plata: Argentina.

Gobierno de la República, 2012, *Ley General de Protección Civil*, Mexico D.F.: Diario Oficial de la Federación, México.

Godínez Guevara, L., and Lazos Chavero E., 2003, "Sentir y percepción de las mujeres sobre el deterioro ambiental: Retos para su empoderamiento," in Tuñón, E. (coord.), *Género y Medio Ambiente*, Mexico City: Ecosur, SEMARNAT and Plaza y Valdés, 145–178.

Ímaz, M., Blázquez, N., Castañeda, I., Chao, V., Beristain, A., and Mussot, L. (coords.), 2014, *Cambio climático, miradas de género*, México City: UNAM y PNUD.

INEGI (Instituto Nacional de Estadística y Geografía), 2010, *Censo de Población y Vivienda*. Retrieved August 12, 2019, from www.inegi.org.mx/programas/ccppv/2010/?ps=microdatos

IPCC (Parry, M.L., Canziani, O.F., Palutikof, J.P., van der Linden, P.J., and Hanson, C.E., eds.), 2007, *Assessing key vulnerabilities and the risk from climate change. Climate Change 2007: Impacts, Adaptation and Vulnerability. Contribution of Working Group II to the Fourth Assessment Report of the Intergovernmental Panel on Climate Change*, Cambridge: Cambridge University Press, 779–810.

IPCC (Field, C.B., Barros, V.R., Dokken, D.J., Mach, K.J., Mastrandrea, M.D., Bilir, T.E., Chatterjee, M., Ebi, K.L., Estrada, Y.O., Genova, R.C., Girma, B., Kissel, E.S., Levy, A.N., MacCracken, S., Mastrandrea, P.R., and White, L.L., eds.), 2014, *Climate Change 2014: Synthesis Report Summary for Policymakers*, Geneva, World Meteorological Organization, www.ipcc.ch/pdf/assessment-report/ar5/syr/AR5_SYR_FINAL_SPM.pdf.

Jáuregui, E., 2003, "Climatology of landfalling hurricanes and tropical storms in Mexico." *Atmósfera*, 16, 193–204.

Juran, L., and Trivedi, J., 2015, "Women, gender norms, and natural disasters in Bangladesh," *Geographical Review*, 105(4), 601–611.

Kabeer, N., 1994, *Reversed Realities. Gender Hierarchies in Development Thought*. London and New York: Verso, p. 350.

Lazos Chavero, E., and Mirenda, C., 2021, "La seguridad alimentaria en la política mexicana de cambio climático. Desigualdades desde la perspectiva de género e indígena," in Solorio I. (ed.), *México ante la encrucijada de la gobernanza climática*, Facultad de Ciencias Políticas y Sociales, UNAM, 121–128.

Leach, M., Mehta, L., and Prabhakaran, P., 2016, "Sustainable development: A gendered pathways approach," in Leach, M., (ed.), *Gender Equality and Sustainable Development*, Oxon, New York: Routledge, 1–33.

Lucatello S., 2008, "Cooperación internacional para el desarrollo: pobreza y prevención de desastres, reflexiones para la política pública," in Lucatello, S., D. Rodríguez Velázquez, and M. Garza Salinas (coords.), *Políticas públicas y desastres*, Mexico City: Instituto Mora, 63–79.

Maass, M., Ahedo-Hernándeza R., Araiza S., Verduzco A., Martînez-Yrîzar A., Jaramillo V. J., et al., 2018, "Long-term (33 years) rainfall and runoff dynamics in a tropical dry forest ecosystem in western Mexico: Management implications under extreme hydrometeorological events," *Forest Ecology and Management*, 426, 7–17 http://dx.doi.org/10.1016/j.foreco.2017.09.040

MacGregor, S., 2010, "'Gender and climate change': from impacts to discourses," *Journal of the Indian Ocean Region*, 6(2), 223–238.

Marchezini V., 2014, "La producción silenciada de los 'desastres naturales' en catástrofes sociales," *Revista Mexicana de Sociología*, 76(2), 253–285.

Mirenda, C., and Lazos-Chavero E., 2021, "Cultural vulnerability, risk reduction and gender equity: two Mexican coastal communities," *Environmental Hazards*. DOI: 10.1080/17477891.2021.1945996

Mondragón Herrada, C., 2012, "El avance de los derechos de las mujeres en México," in M. Moreno Bonett and R. M. Álvarez González (eds.), *El Estado Laico y los Derechos Humanos en México: 1810-210, Volume II, Biblioteca Jurídica Virtual, Repositorio Universitario, Instituto de Investigaciones Jurídicas, UNAM*, 193–208. http://ru.juridicas.unam.mx:80/xmlui/handle/123456789/32564

Neumayer, E., and Plümper T., 2007, "The Gendered Nature of Natural Disasters: The Impact of Catastrophic Events on the Gender Gap in Life Expectancy, 1981–2002," *Annals of the Association of American Geographers*, 97(3), 551–566. DOI: 10.1111/j.1467-8306.2007.00563.x https://doi.org/10.1111/j.1467-8306.2007.00563.x

Oropeza Orozco, O., 2008, "Prevención de desastres de origen hidrometeorológico, una prioridad nacional: El caso de las inundaciones," in Lucatello, S., D. Rodríguez Velázquez y M. Garza S. (coords.), *Políticas públicas y desastres*, Mexico: Instituto Mora, 125–144.

Pérez Castañeda, J.C., and Mackinlay, H., 2015, "¿Existe aún la propiedad social agraria en México?," *Polis: Investigación y Análisis Sociopolítico y Psicosocial*, Universidad Autónoma Metropolitana Unidad Iztapalapa México, 11(1), 45–82.

Rai, S., and Waylen, G. (eds.), 2013, *New Frontiers in Feminist Political Economy*. London: Routledge.

Red de Género y Medio Ambiente, 2010, *Aportes de las experiencias comunitarias a las estrategias de adaptación al cambio climático en México desde una perspectiva de género*, GWA, IMTA, AECID, Gobernment of Spain.

Registro Agrario Nacional, 2020, "Sujetos de núcleos agrarios certificados y no certificados," www.ran.gob.mx/ran/indic_gen/nucag-certynocert-resultados-2019.pdf

Ribot, J. C., 2011, "Vulnerability before adaptation: toward transformative climate action," *Global Environmental Change*, 21(4), 1160–1162.

Richard, A., 2008, "Withered Milpas: Governmental Disaster and the Mexican Countryside," *Journal of Latin American and Caribbean Anthropology*, 13(2), 387–413.

Rocheleau, D., Thomas-Slayter, B., and Wangari, E. (eds.), 1996, *Feminist Political Ecology. Global Issues and Local Experiences*. London: Routledge.

Rodríguez Velázquez D., 2008, "La política pública frente a desastres en el contexto de la reforma del Estado. Opciones desde la sociedad civil," in S. Lucatello, D. Rodríguez Velázquez y M. Garza S. (coords.), *Políticas públicas y desastres*, México: Instituto Mora, Red Mexicana de Estudios Interdisciplinarios para la Prevención de Desastres, pp. 39–62.

Rodríguez Velázquez, D., 2011, "Adaptación y prevención de desastres en el contexto nacional de inseguridad," in S. Lucatello, and D. Rodríguez Velázquez (coords.), *Las dimensiones sociales del cambio climático.* Mexico City: Instituto Mora/ENTS- UNAM, 112–158.

Skinner, E., 2011, *Gender and Climate Change. Overview Report*, Institute of Development Studies.

Soares, D., and Murillo-Licea, D., 2013, "Gestión de riesgo de desastres, género y cambio climático. Percepciones sociales en Yucatán, México," *Cuadernos de Desarrollo Rural*, 10(72), 181–199.

Taylor, M., 2015, *The Political Ecology of Climate Change Adaptation. Livelihoods, Agrarian Change and the Conflicts of Development*. Oxon, New York, Routledge.

Vázquez García, V., 2016, "Género y derechos agrarios en México. Reflexiones a más de dos décadas de la reforma de 1992," in M. Velázquez Gutiérrez, V. Vázquez García and A. de Luca Zuria (coords.), *Transformaciones ambientales e igualdad de género en América Latina. Temas emergentes, estrategias y acciones*, Cuernavaca, Universidad Nacional Autónoma de México, Centro Regional de Investigaciones Multidisciplinarias, 173–195.

Vázquez García, V., 2017, "Género y cambio climático: reflexiones desde la teoría feminista," *Defensor*, 5, 43–46.

Vázquez García, V., Castañeda Camey, I., Molina Rosales, D., Sosa Capistrán, D. M., Chablé Can, E. M. and del Rivero Castañeda, L., 2015, "Género y cambio climático. Estado del arte y agenda de investigación en México," in C. Gay and García, A. Cos Gutiérrez, and T. Peña Ledón (eds.), *Reporte mexicano de cambio climático. Grupo II. Impactos, Vulnerabilidad y Adaptación*, Mexico city: Programa de Investigación en Cambio Climático, Universidad Nacional Autónoma de México, 315–329.

Warman, A., 2003, *La reforma agraria mexicana: una visión de largo plazo*, in FAO, Land Reform, Land Settlement and Cooperatives, 2003/2, Economic and Social Development Department, available at www.fao.org/3/j0415t/j0415t09.htm.

Wilches Chaux, G., 1993, "La vulnerabilidad global," in A. Maskrey (ed.), *Los desastres no son naturales*, La Red – Red de Estudios Sociales en Prevención de Desastres en América Latina, 11–44.

Wisner, B., Blaikie P., Cannon T., and Davis I., 2003, *At Risk: Natural Hazard, People s Vulnerability and Disaster. Second Edition*. London: Routledge.

Young, K., Wolkowitz, C., and McCullagh, R., 1984, *Of Marriage and the Market: Women's Subordination Internationally and Its Lessons*. London: Routledge & Kegan Paul.

PART II
Education

6

SUSTAINABLE GENDER EQUALITY

Opening the black-box of quality assessment in higher arts education

Lilli Mittner, Rikke Gürgens Gjærum, and Hilde Synnøve Blix

6.1 Introduction

Issues of gender equality (GE) in academia receive increasing attention in terms of social justice and human resource management, and quality of research (Leach, 2016; Thompson, 2018; Pandey, 2019). While there is a body of research on gender representations and measures for women's advancement in academia (Riegraf et al., 2010), little research has been conducted on the intertwinement of quality assessment and gendered mechanisms in a sustainable perspective. The starting point for our discussion of short empirical examples is the assumption that insights from the arts and humanities will set quality assessment in academia and beyond into a broader context (Benneworth et al., 2016). With our chapter, we are 'opening a black box' (Lamont, 2009) of quality assessment in higher arts education in Norway and scrutinize in what ways current practices may bring academia towards more sustainable working environment (or not).

The Sustainable Development Goals (SDGs) are the worlds action plan to eliminate poverty and inequality and to stop the climate change before 2030. The 17 SDGs are grouped into five subgroups: people, planet, prosperity, peace, and partnerships. In this chapter we point out three particularly important goals that has great impact on higher education and research: SDG 4: Good education, SDG 5: Gender equality, and SDG 9: Innovation and infrastructure. Since 2000, there has been a progress in achieving the target of universal primary education in total.[1] UNICEF shows that more girls are in school than ever before, and we know that 'investing in girls' education transforms communities, countries and the entire world. /.../ They earn higher incomes, participate in the decisions that most affect them, and build better futures for themselves and their families'.[2] One of the most powerful and proven vehicles for sustainable development is inclusive and good

DOI: 10.4324/9781003174257-8

education. The SDGs focus on the importance of eliminating all forms of discrimination, and one way to achieve this is innovations and infrastructure. R ebuilding infrastructure are key to finding lasting solutions that contribute to long-lasting change. GE in higher education and research are important to promote due to the SDG of building a fairer world where everyone has the same opportunity to contribute, create, fulfil their dreams and use their talents, and most of all to develop a more sustainable world in partnership with other human beings.

6.1.1 *The problem*

Criticism for the lack of gender balance in public institutions such as concert halls, festivals, or museums is often dismissed with a reference to the idea that this is not a matter of gender but a matter of quality. Artists are, so to speak, not selected because of their gender but solely based on the quality of their work. In all art forms, quality assessments are an important part of the artist's everyday life and work. Artists undergo quality assessments in connection with the allocation of funds, access to viewing venues, selection for participation in festivals/biennials, and in the awarding of scholarships or positions.

In academia, artists are assessed in terms of (1) international standard, (2) relevant breadth and depth, and (3) pedagogic competence when competing for positions or seeking promotion (Lovdata, 2006). Quality in the arts is under continuous debate (Biggs & Karlsson, 2010; Mortensen, 2015) and gender issues gained increased attention in the field during the past decades (Nyamko & Twengström, 2009; Ahlqvist et al., 2013, Onsrud et al., 2021, Mittner & Blix, 2021). Committee members who are asked to fairly assess their peers are operationalizing the regulations based on their disciplinary standpoint that can connect to different methodologies, epistemologies, and sets of academic values (Lamont, 2009). Hence, in the moment quality is discussed, power and bias are at work. This can result in conflicting operationalizations of the regulations. The purpose of this chapter is to contribute to discussions of the concept of quality in the arts, to explore in what ways gender matters for how quality is assessed within higher arts education, and to relate the discussion to the SDGs 4, 5, and 9.

6.1.2 *Structure*

The chapter is structured as followed: 1. We present the concept of quality in the arts and show how it connects to power and bias. 2. We then give a short overview over our methods for data collection conducted between 2015 and 2018 as part of the research and intervention project 'Gender Balance in Arts Education' at UiT The Arctic University of Norway. 3. We discuss our results by presenting short empirical examples and relate those to the SDGs 4, 5, and 9. 4. We conclude with an invitation for assessment committees to collectively reflect upon a checklist before starting the assessment process.

6.2 Theory

6.2.1 Quality

The concept of quality is under continuous re-conceptualization within the performing arts (Lund et al., 2001; Eliassen & Prytz, 2016). Little research has been conducted on how quality criteria are defined in recruitment processes. When artistic quality is assessed in academic contexts, different value systems meet. For example, for academic quality (a.k.a. excellence), the number of research papers is of high relevance. For artistic quality (a.k.a. originality), however, what matters most is rather recognition and visibility. What constitutes originality is, however, unclear, precisely because it cannot be defined in advance (Mortensen, 2015, p. 2). Therefore, what is expected in an application for promotion in higher arts education may appear unclear to the applicant when reading the regulations. Assessments in recruitment and promotion processes for artistic positions are based on other criteria than in more traditional scientific positions. According to the Regulations on employment and promotion in teaching and research posts § 1–2 criteria for promotion to professor of academic subjects are for example: 'Scientific level in accordance with established international or national standards' (Lovdata, 2006). Criteria for promotion to the position of professor of performing arts read: 'Comprehensive artistic activity at the highest level according to international standard and relevant breadth and depth in the discipline or discipline at the highest level' (Lovdata, 2006). The quality criteria in traditional academic subjects are in many ways clearer and more predictable than in the arts, because established standards refer to peer-reviewed articles that follow relatively established norms.[3] Concretizing what artistic qualities are is perceived as difficult to express verbally, which often leads to factors such as recognition, resource access, and reputation being used as quality indicators (Stavrum, 2014).

Quality is never absolute but situated and relational and must be operationalized (Eliassen & Prytz, 2016) which takes place in connection to various discourses (Mullany, 2007). For the applicants, it can therefore be challenging to interpreted and understand what the specific expectations for a position are. Within male-dominated working environments, this may be challenging for women who might not be familiar with or part of the discourses that are at stake. Moreover, women tend to underestimate their own competence and status in the application process (Beyer, 1990).

6.2.2 Power

Processes of quality attribution are highly entangled with power, status, elites, and prestige (Ginsburgh, 2003; Abelsen, 2009; Stavrum, 2014). Hence, recognition and visibility are important for artists' career development (Blix, 2017; Heian, 2018). Looking closer how actors in the art field refer to different forms of symbolic power (Bourdieu, 2007) enables us to unvail normative complexities and make implicit assessment practices explicit for those who work in higher arts education and beyond. Brink (2013), for example, describes how individual attitudes, taste, judgement, and organizational-structural routines control specific selection mechanisms.

Brandser and Sümer (2016) refer to the ways in which organizational processes can have different effects on women and men's career development in academia. In some contexts, women's career paths are described as *labyrinths* (Eagly & Carli, 2007) as there can be more obstacles and wrong turns along the tracks of a female academic. In addition, the emphasis on excellence together with neoliberal thinking has become more dominant in academia in recent years by initiatives such as the Centers for Outstanding Research or the EU Framework Program for Research and Innovation, with its emphasis on 'Excellent Science'. This kind of excellence focus has also traditionally been part of artistic quality assessment (Rønning, 2012), master narratives, star culture, and the idea of the (male) genius (Elkins, 2013). This further connects to the various dispositions of white, masculine hegemony that can prevent both women and others who are marginalized in academic cultures to reach top positions in academic fields (Nielsen, 2016).

6.2.3 Bias

Implicit biases are subconsciously cognitive functions that may lead to errors of judgement (Holroyd et al., 2017). Research on academic practices has revealed a long list of implicit biases connected to gender, assessment of research projects, and also research methodologies or disciplines (Beyer, 1990; Gouldin & Rose, 2000; Biernat et al., 2012; Moss-Racusin et al., 2012; Rossiter, 2012). The National Research Ethics Committees in Norway point out that:

> The researcher's own attitudes and awareness of opportunities for his own bias and research bias is an important prerequisite for reducing the risk of bias as well as reducing any negative effects as a result of research bias.[8]

Both explicit and implicit gender bias connects to gender stereotypes that are expressed in attitudes, the way people speak and act (Reuben et al, 2014). Despite over 100 years of women's movement, we have still empirical evidence based on data from the Scandinavian countries that women's work is statistically considered to be weaker than men's (Wennerås & Wold, 1997). In the international research community, when an application carries a male name, it has a much higher chance of being considered positive than if the application carries a female name, even though the applications are identical (Moss-Racusin et al., 2012). The experimental design of Moss-Racusin et al. (2012) has been recently applied for the Nordic region and the gender bias as key explanation of the persistent gender gap in academia has been contested (Carlsson et al., 2020). Still, women in management positions are more often at risk of being negatively assessed regardless of whether they choose to follow expectations of feminine behaviour (collaborative) or show a high degree of competitive behaviour (Crawford, 1995; Mullany, 2007; Lakoff, 2008). There are further greater income disparities for male and female artists than in other professional occupations, and one explanation model points at the combination of a gendered division of labour and informal assessment processes characterized by hegemonic understandings of quality (Heian, 2018). Heian concludes that it is

> important to gain more systematic knowledge of the type of competence valued in female versus male artists, the degree to which the skills are used differently, and the mechanisms that apply to gendered recruitment to different styles, genres and positions in the art field.
>
> *Heian, 2018, p. 64*

Finally, what seems to be significant for the arts and humanities is the long-lasting idea that creative thinking connects to masculinity (Proudfoot et al., 2015).

6.3 Methods

The intervention and research project 'Gender Balance in Arts Education' was conducted at the Faculty of Fine Arts at UiT The Arctic University of Norway from 2015 to 2018. The research component of the project aimed (1) to map the gender balance in the field, (2) to identify GE strategies, and (3) to analyse institutional attitudes towards the deployment of these strategies.

In the past 25 years, approximately the same number of female and male students have completed higher arts education in Norway. Still, men occupy most of the leading positions in art arenas both inside and outside academia, and the lack of gender balance in the art field proves to be persistence (Abelsen, 2009; Gjærum et al., 2015). In 2017, for example, there were 32% women at the professor/lecturer level in art education in Norway, and in the music subjects at universities and scientific colleges, the proportion of women was particularly low with only 21% female professors/lecturers (Gunnes & Olsen, 2018).

We wanted to investigate possible reasons for these statistics provided by NIFU (Gunnes & Olsen, 2018) and collected 24 committee reviews and conducted five semi-structured in-depth interviews (Kvale & Brinckman, 2009) in the form of intimate conversations (Strom, 2018) with strategically selected experienced members and leaders of assessment committees in the field of higher arts education in Norway in 2015. We asked how quality is understood and regulated in the assessment of top-level competence in higher arts education, and to what extent and in what ways committee members acknowledge gender bias as significant in the ways quality is acknowledged and described.

The sample of informants in the study consists of four professors and a lecturer with long experience as members and leaders of expert committees. The informants, three women and two men aged 58–76, have been full-time employees of various Norwegian universities and colleges for several years in the fields of music, drama, and visual arts. Because the individual art fields are small and characterized by close ties of knowledge, for reasons of anonymity, we chose not to disclose which informants represent the individual field. The informants are given fictitious names in the article.

As a part of the study, we collected 24 assessment reports for professorship promotions in Norwegian arts education. We analysed collectively (Eggebø, 2020) the language and arguments for acceptance (or not) regarding these employees' applications for full professorships. We work within the field of higher arts education and we were involved in data collection (HB & RG) and data transcription (HB & LM). We have been ourselves in the role of assessing their peers, and RG has been part of one of the assessment committees we are referring to in this chapter. Being part of the normative complexities that we are studying was central for our research design and gave unique access to data collection in terms of trust and relational ethics (Becker, 2012).

6.4 Discussion of results

We will now invite the reader into the Norwegian arena of higher arts education by presenting short empirical examples that give insight into the way quality is assessed in the light of the SDGs 4, 5, and 9. The examples derive from a collective thematic analysis of 5 intimate conversations with experienced committee members and 24 assessment reports that are differing in design, length, and argumentation style.

6.4.1 SDG 5: Gender equality

In Scandinavian countries, as everywhere else, women and men continue to face different opportunities in the labour market and in the workplace; the labour market is highly gender segregated and women are underrepresented in top decision-making positions (Holgersson & Wahl, 2017). The SDG 5 seeks to 'ensure women's full and effective participation and equal opportunities for leadership at all levels of decision-making'. One measure to reach this goal is to assure women access to top-level positions in higher education. In our material, we found that gender is not an explicit issue in competence assessment to professorship positions; however, both the way applicants advocate for quality and the way committee members assesses quality is highly gendered.

When asked directly, committee members replied that gender is not an issue when evaluating applications for full professorships. Gender was further not mentioned explicitly in any of the assessment documents. When we invited our informants to take gender perspectives into account, none of them connected gender to power and discrimination. The following quote document how gender is rather understood in terms of diversity:

> When it comes to gender, I'm sure there are differences, but I think most people in my field are interested in balancing those as much as possible.

> [...] We are only few colleagues, so it is quite easy to get an imbalance. And I have been involved many times in the discussions about whether gender plays a role or not. It is said: 'Well, we just want the best ones', so we do not need to think about gender. I think you must think about it. It is often difficult to separate the two top-listed [...] I think that if you have an imbalance, then you can think once more. People easily identify with their own gender, or someone who looks like yourself. This can apply to both gender and other attributes. It is important to consider diversity within your team.
>
> *informant 3*

The notion of gender balance and diversity connects to current policy recommendations in Scandinavia which becomes exemplary visible in the change of name from 'Committee for Mainstreaming – Women in science' (2004) via 'Committee for Gender Balance in Research' (2010) to 'Committee for Gender Balance and Diversity in Research' (2014). None of our informants or committee reports addresses gender bias and its effects on assessment. However, a variety of references were made in connection with gender stereotypes, such as for example:

> If a man uses elbows, it may be correct, but if a woman does the same, she becomes dangerous, because she is strong. It wasn't that way they framed it, but it is the reality.[4]
>
> *informant 4*

This way of framing gender perspectives is surprising and points at a lack of reflection within the board room (Seierstad et al., 2017) and lack of awareness among decision-makers (Wahl, 1995). What is needed is a culture in the board room in which committee members feel secure to address their own biases on gender stereotypes without perpetuating blindly what is perceived as 'reality'.

Further, gender is not just at work when competence is assessed but also in the way it is presented. We are not able to generalize from our material, but we want to point out two observations: both teaching responsibilities and merits as well as international activities are presented to varying extend by the applicants. One applicant was promoted with only small documentation of his merits as pedagogue which points at the fact that teaching was not explicitly part of the committee's assessment. This is surprising when taking into account that there are specific positions in Norway which exclusively focus on research while professorship positions are dedicated 50% research and 50% teaching. We further observed large differences in the extent to which international activities and networks are presented by the applicants and assessed by the committees. Since both education and mobility are highly gendered (Nordiska ministerråd, 2010), the committee's latitude can be a threat for those who apply for top positions.

6.4.2 SDG 4: Good education

For achieving a sustainable diversity of practices, perspectives, and role-models within higher education, we need gender balance in the group of those who are teaching students, but also a more explicit discussion of what 'good quality' entails in higher education at all. In our material, the level of required pedagogical competencies of those who seek professorship positions is seldom emphasized, and scarcely accurate described. We further found a discrepancy between the assessment boards' desire to emphasize teaching skills, and the will to forefront this in an assessment of professor competency. The following example derives from the assessment of a male applicant who documented his teaching merits but who was not assessed competent for professorship.

> His pedagogical documentation forms a very strong part of the application. He refers to several of his former students, who today hold [...] positions in leading orchestras. The committee would like to pay tribute to his great pedagogical commitment. Both as an instrumental teacher, initiator to masterclasses with renowned educators, and organizer of study trips.[5]
>
> *assessment document 4*

This case indicates that teaching is not equally part of competence assessment for professorship positions in higher art education as research activities. What rather counts are artistic originality that is linked to acknowledgement and visibility (Battersby, 1989). One of our male informants pointed out:

> I believe that the quality of teaching should also have a high level, because, I just saw, [...] we brought in some 'chieftains', however they just communicate with themselves, so this had little value [for our students]. They can play so well, but we see how necessary it is to have the best teachers. When the level is very high anyway, it is of great value how they teach.[6]
>
> *informant 4*

This quote points out that standards of academic practices and excellence are not necessarily associated with excellent teaching and collaborative work with students. This becomes problematic when standards in higher education are defined 'male standards' (Noddings, 1991).

However, when conversing with our informants, we found tendencies towards counter-discourses that challenge hegemonic thinking practices in the field. An increased focus on educational merits, as the next example demonstrates, points towards a shift of what constitute excellence (Heine-Bennet et al., 2020).

> [He] has chosen to not explicitly document his practical-pedagogical competence, beyond the reference to his formal competence within teacher

> education. The committee considers this as unfortunate since possible links between his artistic practice and teaching have not been clarified.[7]
>
> *assessment document 10*

What is needed are committees that consider missing pedagogic merits as 'unfortunate' and that operationalize the legal requirements by setting more explicit standards for what is meant by 'pedagogic competence'. If pedagogic practices are valued in promotion for professorship, then more people who are experts in teaching-based research might enter universities and become role models for our students and change-makers towards more sustainable academic practices.

6.4.3 SDG 9: Innovation and infrastructure

According to the Norwegian working group who just published the report 'SDG – Quality in higher education: Developing a platform for sharing of ideas and practices within the universities': We need more innovative, multisectoral, and multithematic support systems and partnerships that can be brought to scale quickly, engaging all parts of society, and particularly the young people who will be key to the success of the 2030 Agenda (University of Bergen Working Group, 2020, p. 20).

Let's dive into the terrain of competence assessment and gender in higher education and research in the perspective of *building a more resilient infrastructure through innovation*. Based on our data, we claim that building such an infrastructure can be done along three pathways: (1) more predictable assessment criteria, (2) meta-discussions on practices of academic assessment, and (3) formal training of (new) committee members.

First, we find in the conversations with committee members that they experience the Norwegian regulation for evaluating professorship, in LOV-2005-04-01-15-§6-3, too vague. However, clearer redefinition of the regulation with specific measurements would contribute to less gendered assessment (Blix et al., 2019). One of the committee members told us that:

Those who are in the field develop a "nose" for what's good enough.[8] (informant 3)

And another one pointed out:

> But the committee is a combination of different people in power positions, so you need to negotiate, sometimes its hopeless because there often is a gatekeeper who doesn't want to give up.[9]
>
> *informant 2*

The informants talk about 'bias' and sometimes 'random processes' shaped by subjective interpretations of the operative definition of artistic 'quality' in an application for professorship. 'Gut-feeling and likes and dislikes' is mentioned as the ways to explain how quality is understood by some committee members. It is further mentioned that some female applicants tend to find both the *lack of language* to describe their academic experience and artistic research verbally in the application, and *vague criteria for promotion*, as barriers in an assessment culture still dominated by men.

Second, we find a track in our data that indicates a need for more meta-discussions among committee members on practices of academic assessment in the arts. One example from the assessment documents (2013) shows an interesting aspect of the problem in defining quality in an international context:

> 'There is one key term that is difficult to operationalize when it comes to artistic competence assessment for professorship positions, and this is 'artistic practice on top level on international standard'. It is as problematic to differentiate between 'high level' and 'top level', as it is to differentiate between 'international standard' and something else ('national' standard?). A strict operationalization of 'top international level' will always imply a small group of artists with a field. However, the committee decides not to operationalize the term so strict. The committee neither starts out from the assumption that 'international standard' that is higher than 'national standard' but rather 'international standard' means that the artworks can have international relevance or gain international interest'.
>
> *assessment document 16*

We see that the committee choose to justify their interpretation of one sentence from the regulation LOV-2005-04-01-15-§6-3 in the assessment document: 'Extensive artistic activities at the highest-level conforming to international standards'. We find several similar examples in the rest of the assessment documents. Especially, the terms such as 'relevant breadth', 'highest level', and 'international standard' are problematized by the committee members, both in the five transcribed conversations and in the assessment documents.

Third, we discovered a need for a proper formal training of (new) committee members; thus, the senior members strongly disagree on several aspects in the assessment processes. In the conversations, committee members claimed that women tend to: 'underestimate their own competence' and 'female committee members is much easier to work with – they are prepared and structured', but in the end, 'women struggle to take a decision and conclude'. We find in the data a need for a proper formal training in assessment of academic and artistic quality to ensure equal assessment.

One aspect that we find as a challenge for the committee members when they evaluate applications for professorship in the Arts are different expectations, e.g. towards to use of theory. One assessment document raises the question of the lack of theory in the application:

> Experience from many years of research and development activity are documented, among other things were referring to a well-developed teaching method for conceptualising dramaturgical work. The committee find this work to be interesting and important. The problem in the existing documentation is that even if the basic idea of this teaching is easily interpreted from the attached

> teaching plans, it is not possible to create an impression of the theoretical side of the conceptual work. How are the chosen terms on concepts presented, analysed, and placed in connection with a theory about the creative work?
>
> *assessment document 16*

The three pathways exemplified above (criteria, meta-discussions, and formal training) indicate possible ways out of 'the assessment vagueness problem'. Based on our data we suggest that a new framework for assessment is possible to create as a future innovative infrastructure in academic assessment in line with the SDG 9. Innovation is often understood as change that is meaningful. Thereby an innovative infrastructure for academic assessment must come to use in a transformative way; a more democratic and transparent way than today's practice. Thus be meaningful for both appliers for professorship and committee members. International research report that gender balance in academia depends on cultural and system related factors (Søndergaard & Hæjgaard, 2003; Nielsen, 2016). We can work towards the SDG 9 in higher education and research by raising awareness of the significance of re-building a resilient infrastructure as open and equal for all applicants so that they now before they apply which criteria will be used.

6.5 Summing up

The overall aim of this chapter is to contribute to more transparent discussion of quality higher arts education. To create 'a more sustainable world, individuals and communities must become sustainability change-makers' (University of Bergen Working Group, 2020). This means that we need to find ways to become part of the change we wish for. In this chapter, we throw light on how gender matters for how quality is assessed within higher arts education in Norway and how practices of competence assessment can be understood in a sustainability goal perspective. What the theories present, and our field examples indicate, is a need for increased exploration on how academia can contribute to more transparent discussions of quality, gender, and competence.

Our interpretation of the data material calls for: (1) more transparent, explicit assessment practices and (2) a clearer definition on what is meant by quality. We argue that more transparent and explicit quality criteria could result in a more sustainable application procedure. This means that all applicants are equally informed about the requirements for academic, artistic, and pedagogical competence *before* they apply for a position or promotion. Within a more explicit regulation, it won't be possible for committee members to allow their own taste, personal judgement, and cultural preferences to govern the assessments (Lamont, 2009). Hence, an implication of our study is that a new and improved regulation should be drawn up, indicating what is meant by relevant breadth and depth or scientific level in accordance with international standards. Applicants of all genders need to know how many (and what) articles, performances, or classes need to be submitted, organized, or taught to be considered for professorship and excellence in a subject. One of the greatest obstacles in approaching the interconnected sustainable challenges today:

> is silo-thinking and a potential inability to overcome internal and institutional barriers towards solving the SDGs. Silo-thinking is a hinderance for transdisciplinary collaboration within universities, typically reflecting thematic, organizational, and financial structures, which again are promoted by governmental financing systems, research funding organizations, and approved publication channels rooted in scientific culture.
>
> *University of Bergen Working Group, 2020*

Continued silo-thinking and keeping quality markers unclear might contribute to maintaining a not-sustainable gender imbalance in academia. In addition, we suggest the following checklist for assessment committees to collectively reflect upon before starting the assessment (Figure 6.1).

To sum up, we need national arenas for critical meta-discussions about what constitutes *quality* to fulfil the SDG aims of good education, and GE through innovation and infrastructure. Those arenas must be implemented in the life-long learning education strategies in the universities[10] and through committee work, academic- and artistic practices.

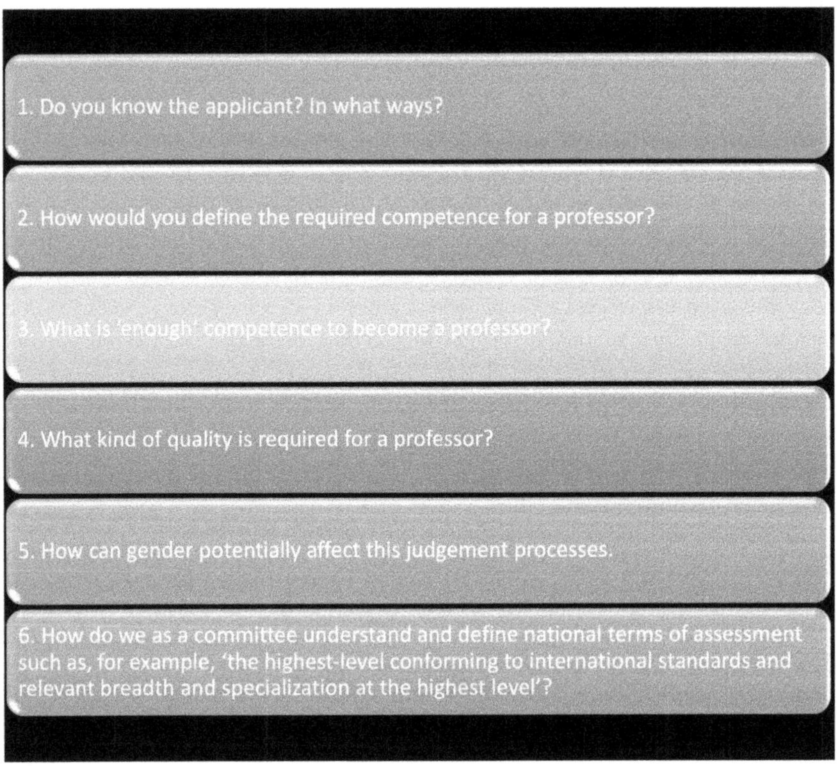

FIGURE 6.1 Checklist for assessment committees.

Acknowledgements

We would like to thank our informants who are experts in competence assessment in the field of higher arts education in Norway. Their contribution is central to the study and will, hopefully, enhance assessment practices in the field towards more sustainable education, gender relation and innovation. We further thank the Research Council of Norway for funding our research as part of the research and intervention projects 'Gender Balance in Art Education' (RCN 2015-2018/245487) and 'Prestige. Gender Balance in Research Leadership' (RCN 2018-2021/282862).

Notes

1 Goal 4: Quality education (undp.org).
2 Girls' education | UNICEF.
3 https://dbh.nsd.uib.no/publishing.
4 "Hvis en mann bruker albuer så er det liksom korrekt, og det gjør man, men hvis et kvinnfolk gjør det så er hun farlig, for da er hun sterk. Det var jo ikke slik de sa det, men det er jo det som er realiteten."
5 "Hans pedagogiske dokumentasjon utgjør en meget sterk del av søknaden. Han viser til en rekke av sine tidligere studenter, som i dag har [...] stillinger i ledende orkestre. Komiteen vil gjerne gi honnør til hans store pedagogiske engasjement. Både som instrumentallærer, initiativtager til mesterklasse med anerkjente pedagoger og organisator av studiereiser."
6 "Jeg mener jo at kvaliteten på undervisningen også må ha et høyt nivå, for jeg ser jo, [...] vi hentet inn noen høvdinger, men de står og kommuniserer kun til seg selv, så det har så lite verdi. Da kan de spille så godt som de vil, men vi ser hvor nødvendig det er å ha de beste underviserne. Når det uansett er veldig høyt nivå, så er det av stor verdi hvordan de underviser."
7 "[Han] har valgt ikke direkte å dokumentere sin praktisk-pedagogiske kompetanse, utover henvisningen til sin formelle kompetanse via lærerutdanning. Komiteen vurderer dette som uheldig, da mulige koblinger mellom hans kunstneriske praksis og undervisning ikke er tydeliggjort."
8 "Men de som er i feltet utvikler litt sånn «nese» for hva som skal til. Og der kommer vi litt inn på kjønn. Når man sitter i et rom sammen med andre og ser på 20 kandidater, så merker man i rommet hvordan stemningen er."
9 "Da er det folk som har posisjoner og. så du må virkelig drive å forhandle fram og tilbake og noen ganger så er det håpløst, og andre ganger er det mulig. Men det er alltid en eller annen sånn tempelvokter som sitter og ikke vil gi seg på visse ting."
10 Collection of Lifelong Learning Policies and Strategies | UIL (unesco.org).

References

Abelsen, M. (2009). *Kvinnelige kunstneres symbolske kapital – Om kunstinstitusjonens ideologiske maktstruktur*. Institutt for kunst-og medievitenskap NTNU. http://hdl.handle.net/11250/243225

Ahlqvist, V., Andersson, J., Berg, C. H., Kolm, C., Söderqvist, L., & Tumpane, J. (2013). *Jämställdhetsobservationer i ett urval av vetenskapsrådets beredningsgrupper 2012 = Observations on gender equality in a selection of the Swedish research council's evaluation panels*. Vetenskapsrådet.

Battersby, C. (1989). *Gender and Genius: Towards a Feminist Aesthetics*. London: Women's Press.

Becker, C. U. (2012). The Inherent Ethical Dimension of Sustainability – Toward a Relational Ethical Perspective. In C. U. Becker (Ed.), *Sustainability Ethics and Sustainability Research* (pp. 17–20). Dordrecht: Springer Netherlands. https://doi.org/10.1007/978-94-007-2285-9_3

Benneworth, P., Gulbrandsen, M., & Hazelkorn, E. (2016). *The Impact and Future of Arts and Humanities Research*. London: Palgrave Macmillan UK. https://doi.org/10.1057/978-1-137-40899-0

Beyer, S. (1990). Gender Differences in the Accuracy of Self-evaluations of Performance. *Journal of Personality and Social Psychology*, *59*, 960–970.

Biernat, M., Tocci, M. J., & Williams, J. C. (2012). The Language of Performance Evaluations Gender-Based Shifts in Content and Consistency of Judgment. *Social Psychological and Personality Science*, *3*(2), 186–192. https://doi.org/10.1177/1948550611415693

Biggs, M., & Karlsson, H. (2010). Evaluating Quality in Artistic Research. *The Routledge Companion to Research in the Arts*, pp. 405–424. London: Routledge.

Blix, H. S. (2017). Framsnakk som likestillingstiltak. In *Kilden*. http://kjonnsforskning.no/nb/2017/01/framsnakk-som-likestillingstiltak

Blix, H. S., & Mittner, L. (2018). Balansekunst i utøvende musikkutdanning. En empirisk undersøkelse av lederes synspunkter på kjønnsbalanse og likestillingsarbeid ved de utøvende musikkutdanningene i Norge. [The art of balance in music education]. *Tidsskrift for Kjønnsforskning*. https://doi.org/10.18261/issn.1891-1781-2018-01-02-07

Blix, H. S., Mittner, L., & Gürgens Gjærum, R. (2019). I kjønneste orden? Om definisjon av kvalitet og betydningen av kjønn i vurdering av professorkompetanse. *Nordic Journal of Art and Research*, *8*(1). https://doi.org/10.7577/information.v8i1.2853.

Bourdieu, P. (2007). *The Masculine Dominance* (2nd ed.). Viborg: Times are Changing.

Brandser, G., & Sümer, S. (2016). *Farefull ferd mot toppen. Kjønnsbalanse i forskning og forskningsledelse ved forskningsog utdanningsinstitusjoner i Bergen*. Stein Rokkan senter for flerfaglige samfunnsstudier. www.uib.no/sites/w3.uib.no/files/attachments/rapport_3-2016_brandser_og_sumer_0.pdf

Brink M. van den. (2013). Gender in Academic Networking: The Role of Gatekeepers in Professional Recruitment. *Journal of Management Studies*, *51*(3), 460–492. https://doi.org/10.1111/joms.12060

Carlsson, M., Finseraas, H., Midtbøen, A. H., & Rafnsdóttir, G. L. (2020). Gender Bias in Academic Recruitment? Evidence from a Survey Experiment in the Nordic Region. *European Sociological Review*, *jcaa050*. https://doi.org/10.1093/esr/jcaa050

Crawford, M. (1995). *Talking Difference: On Gender and Language*. London: Thousand Oaks, Calif: SAGE.

Eagly, A. H., & Carli, L. L. (2007). *Through the Maze: The Truth About How Women Become Leaders*. Boston, MA: Harvard Business School Press.

Eggebø, H. (2020). Kollektiv kvalitativ analyse. *Norsk sosiologisk tidsskrift*, *4*(02), 106–122. https://doi.org/10.18261/issn.2535-2512-2020-02-03

Eliassen, K. O., & Prytz, Ø. (Eds.). (2016). *Kvalitetsforståelser: Kvalitetsbegrepet i samtidens kunst og kultur*. Kulturrådet: Fagbokforlaget.

Elkins, J. (2013). *Master Narratives and their Discontents*. Oxford: Taylor and Francis. https://doi.org/10.4324/9780203699713

Fürst, E. L. (1988). *Kvinner i akademia – Inntrengere i en mannskultur?: Om ansettelsesprosessen ved universitet og distriktshøgskoler*. Oslo: NAVFs sekretariat for kvinneforskning.

Ginsburgh, V. (2003). Awards, Success and Aesthetic Quality in the Arts. *The Journal of Economic Perspectives*, *17*(2), 99–111. https://doi.org/10.1257/089533003765888458

Gjærum, R. G., Vollan, I., & Bjerkestrand, K. B. (2015). Levd skuespillerliv og rolletildeling. En studie av kvinner i norsk institusjonsteater. *InFormation – Nordic Journal of Art and Research*, *4*(2), 1–15. https://doi.org/10.7577/if.v4i2.1548

Goldin, C., & Rouse, C. (2000). Orchestrating Impartiality: The Impact of "Blind" Auditions on Female Musicians. *The American Economic Review*, *90*(4), 715–741.

Gunnes, H., & Olsen, B. M. (2016). *Kvinner i kunstfagene*. NIFU. www.nifu.no/publications/1428242/

Gunnes, H., & Olsen, B. M. (2018). *Statistikk om kvinner i kunstfagene 2009–2017* (NIFU-Arbeidsnotat). http://kunstfak.no/wp-content/uploads/2019/05/NIFU_2018_Oppdatert_statistikk_om_kvinner_i_kunstfagene.pdf

Haine-Bennett, E., Bergsieker, H. B., Coe, I. R., Koch-Kraft, A., Langelier, E., Morrison, S., Nikoleyczik, K., Schmader, T., Trivailo, O., Twine, S., & Decker, J. E. (2020). Enacting Workplace Culture Change for Excellence in Research: A Gender Lens. *FACETS*, *5*(1), 228–233. https://doi.org/10.1139/facets-2019-0026

Heian, M. T. (2018). Norske kunstnere og det doble likestillingsparadokset. *Tidsskrift for kjønnsforskning*, *42*(01–02), 47–66. https://doi.org/10.18261/issn.1891-1781-2018-01-02-04

Holgersson, C., & Wahl, A. (2017). Sweden: Work for Change and Political Threats. In C. Seierstad, P. Gabaldon, & H. Mensi-Klarbach (Eds.), *Gender Diversity in the Boardroom: Volume 2: Multiple Approaches Beyond Quotas* (pp. 129–157). Cham: Springer International Publishing. https://doi.org/10.1007/978-3-319-57273-4_6

Holroyd, J., Scaife, R., & Stafford, T. (2017). Responsibility for Implicit Bias. *Philosophy Compass*, *12*(3), n/a-n/a. https://doi.org/10.1111/phc3.12410

Kvale, S., & Brinkmann, S. (2009). *Det kvalitative forskningsintervju*. Oslo: Gyldendal akademisk.

Lakoff, R. (2008). Language, Gender, and Politics: Putting 'Women' and 'Power' in the Same Sentence. In J. Holmes & M. Meyerhoff (Eds.), *The Handbook of Language and Gender* (pp. 160–178). Oxford: John Wiley.

Lamont, M. (2009). *How Professor Think: Inside the Curious World of Academic Judgment*. Cambridge, MA: Harvard University Press.

Leach, M. (Ed.). (2016). *Gender Equality and Sustainable Development*. London: Routledge, Taylor & Francis Group.

Lovdata. (2006). *Regulations Concerning Appointment and Promotion to Teaching and Research Posts*. Norwegian Ministry of Education and Research. https://lovdata.no/dokument/SFE/forskrift/2006-02-09-129#KAPITTEL_2

Lund, C., Mangset, P., & Aamodt, A. (Eds.). (2001). *Kunst, kvalitet og politikk: Rapport fra Norsk kulturråds årskonferanse 2000: Vol. nr 22*. Rådet.

Mittner, L., & Blix, H. S. (2021). Career Paths in Higher Music Education: Challenges for Gender Equality in the Arts. In S.V. Onsrud, H. S. Blix, & I. Lunde Vestad (Eds.), *Gender Issues in Scandinavian Music Education From Stereotypes to Multiple Possibilities* (pp. 180–196). London: Routledge. https://doi.org/10.4324/9781003038207-9

Mortensen, A. (2015). *Teser om kvalitet*. Kulturrådet. www.kulturradet.no/kvalitet/vis/-/teser-om-kvalitet

Moss-Racusin, C. A., Dovidio, J. F., Brescoll, V. L., Graham, M. J., & Handelsman, J. (2012). Science Faculty's Subtle Gender Biases Favor Male Students. *Proceedings of the National Academy of Sciences*, *109*(41), 16474–16479. https://doi.org/10.1073/pnas.1211286109

Mullany, L. (2007). *Gendered Discourse in the Professional Workplace*. Basingstoke; New York: Palgrave Macmillan.

Nielsen, M. W. (2016). Scandinavian Approaches to Gender Equality in Academia: A Comparative Study. *Scandinavian Journal of Educational Research*, 1–24. https://doi.org/10.1080/00313831.2016.1147066

Noddings, N. (1991). The Gender Issue. *Educational Leadership, 49*(4), 65–70. www.ascd.org/ASCD/pdf/journals/ed_lead/el_199112_noddings.pdf

Nordiska ministerrådet. (2010). *Labour Market Mobility in Nordic Welfare States: Vol. 2010:515.* Copenhagen: Nordic Council of Ministers.

Nyamko, S., & Twengström, K. (2009). *On the Way to Equal Performing Arts.* Stockholm: State Culture Council.

Onsrud, S.V., Blix, H. S., & Ingeborg Lunde Vestad. (2021). *Gender Issues in Scandinavian Music Education: From Stereotypes to Multiple Possibilities.* Milton: Taylor and Francis. https://doi.org/10.4324/9781003038207

Pandey, U. C. (2019). *SDG5 Gender Equality and Empowerment of Women and Girls.* Bingley: Emerald Publishing Limited. https://books.emeraldinsight.com/page/detail/SDG5---Gender-Equality-and-Empowerment-of-Women-and-Girls/?k=9781789735246

Proudfoot, D., Kay, A. C., & Koval, C. Z. (2015). A Gender Bias in the Attribution of Creativity: Archival and Experimental Evidence for the Perceived Association Between Masculinity and Creative Thinking. *Psychological Science, 26*(11), 1751–1761. https://doi.org/10.1177/0956797615598739

Reuben, E., Sapienza, P., & Zingales, L. (2014). How Stereotypes Impair Women's Careers in Science. *Proceedings of the National Academy of Sciences, 111*(12), 4403–4408. https://doi.org/10.1073/pnas.1314788111

Riegraf, B., Aulenbacher, B., Kirsch-Auwärter, E., & Müller, U. (Eds.). (2010). *GenderChange in Academia: Re-mapping the Fields of Work, Knowledge, and Politics from a Gender Perspective* (1st edition). Wiesbaden: VS Verlag.

Rossiter, M. W. (2012). A Sophomore Tackles Thomas Kuhn'sStructure of Scientific Revolutions(1964). *Historical Studies in the Natural Sciences, 42*(5), 562–563. https://doi.org/10.1525/hsns.2012.42.5.562

Rønning, A. B. (2012). Kunst, kjønn og estetisk vurdering. *Tidsskrift for kjønnsforskning, 36*(01), 3–17. https://doi.org/10.18261/ISSN1891-1781-2012-01-02

Seierstad, C., Gabaldon, P., & Mensi-Klarbach, H. (2017). *Gender Diversity in the Boardroom: Volume 2: Multiple Approaches Beyond Quotas* (1st ed. 2017.). Cham: Springer International Publishing: Imprint: Palgrave Macmillan.

Stavrum, H. (2014). Hvor mange gullplater henger på veggen? Om danseband og kvalitet. *Sosiologi i dag, 44*(1), 90–116. http://ojs.novus.no/index.php/SID/article/view/846

Strom, K. (2018). *Decentering the Researcher in Intimate Scholarship* (Vol. 31). Bingley: Emerald Publishing Limited. https://books.emeraldinsight.com/page/detail/Decentering-the-Researcher-in-Intimate-Scholarship/?k=9781787546363

Søndergaard, D. M., & Højgaard, L. (2003). *Akademisk tilblivelse: Akademia og dens kønnede befolkning.* København: Akademisk forlag.

Thompson, R. J. (2018). *Leadership and Power in International Development.* Bingley: Emerald Publishing Limited.

University of Bergen Working Group. (2020). *SDG – Quality in higher education: Developing a platform for sharing of ideas and practices within the universities. Report from working group.* www.uib.no/sites/w3.uib.no/files/attachments/sdg_-_quality_in_higher_education_-_report_feb_2020.pdf

Wahl, A. (1995). *Men's Perceptions of Women and Management.* Fritzes. http://urn.kb.se/resolve?urn=urn:nbn:se:kth:diva-61349

Wennerås, C., & Wold, A. (1997). Nepotism and sexism in peer-review. *Nature, 387*(6631), 341–343. https://doi.org/10.1038/387341a0

7

SENSITIZING NEPALESE STUDENTS FOR GENDER MAINSTREAMING IN SUSTAINABLE PLANNING AND DESIGN

Martina Maria Keitsch

7.1 Introduction

Since the past decade, Nepal is experiencing a growing socio-political recognition that sustainable development needs to include gender issues and that development approaches need to be gender-sensitive to be sustainable. Accordingly, the main focus of MSESSD is society-oriented focusing on sustainable development and renewable energy planning and design to improve access to sustainable energy and livelihoods for both women and men, and expand the scale and the quality of sustainable energy initiatives.

The terms "Gender mainstreaming (GM)" and "Gender Equality (GE)" are used synonymously in this chapter. Per definitionem, GE means that women and men have equal rights and opportunities; GM means that gender equality permeates all decision-making levels (Council of Europe 2021). Bjørvik (2017) specifies in the following description:

> Gender Mainstreaming is a globally accepted strategy for promoting gender equality. Mainstreaming involves ensuring that gender perspectives and attention to the goal of gender equality are central to all activities, policy development, research, advocacy/ dialogue, legislation, resource allocation, and planning, implementation and monitoring of programs and projects.
>
> *Bjørvik, 2017*

DOI: 10.4324/9781003174257-9

The Beijing Declaration and Platform for Action (United Nations Women , 1995) adopted by governments at the 1995 Fourth World Conference on Women, established ender mainstreaming as a key element recognizing that:

> Eradication of poverty based on sustained economic growth, social development, environmental protection, and social justice requires the involvement of women in economic and social development, equal opportunities and the full and equal participation of women and men as agents and beneficiaries of people-centered sustainable development.
>
> *ibid., section 16*

In the same line, the Paris Agreement (under the United Nations Framework Convention on Climate Change for the first time) constitutes a Climate Treaty which commits Parties, to promote and consider their respective obligations on human rights, as well as on gender equality, and the empowerment of women (United Nations Climate Change, 2015). Sustainable Development Goal (SDG) 5 (United Nations, 2015) is about achieving gender equality and empowering all women.

In the context of Nepal, the proposed targets for 2030 include eliminating gender disparity in education, wage discrimination at similar work, physical and sexual violence, and all harmful practices, and raising the presence of women in the national parliament (United Nations Development Program, 2021). The achievement of these targets calls for GM in all aspects of development especially in the environment and energy. The promulgation of the new Nepalese Constitution in 2015 (Nepal Law Society et al., 2021) emphasizes regional development and MSESSD assumes renewable energy as significant point of departure for the transformation of the Nepalese society toward equitable livelihoods and gender-fair income generation through regional developments.

7.2 Gender mainstreaming in politics and higher engineering education

7.2.1 Nepal

As in other countries, traditional socio-cultural practices affect development efforts in Nepal, particularly regarding the social and economic exclusion of women, despite politically expressed commitments to eradicate long-standing systems and practices of exclusion. Nepalese Constitution established in 2015 states:

> Ending all forms of discrimination and oppression created by the feudalistic, autocratic, centralized, unitary system of governance, recognizing the multi-ethnic, multi-lingual, multi-religious, multi-cultural and diverse regional characteristics, resolving to build an egalitarian society founded on the proportional inclusive and participatory principles to ensure economic

equality, prosperity and social justice, by eliminating discrimination based on class, caste, region, language, religion and gender and all forms of caste-based untouchability.

ibid., p. 12

Since the beginning of the 21st century, the country has launched various policies, programs, and action plans to promote gender equality and the empowerment of women. The Electoral Knowledge Network (2021) states for example that the Election Act 2008, established a 50% quota for all proportional representation lists (The Electoral Knowledge Network, 2021), . However, GE often remains conceptual and is not translated into reality for women, who are, especially in rural areas, mostly unaware of laws to protect and promote them. Dalit women (see p. 7) and women from indigenous groups face extra difficulties in accessing justice and social services. Many are deprived of citizenship certificates, which results in various difficulties from land ownership to accredited education documents (Bhattachan, 2005). Women's overall participation also remains low at the higher levels of decision-making, in positions of public importance, and in specific sectors. Government records from 2014 show women's representation in public life with around 16% in civil service, 4.5% in the judiciary, 5.8% in the Nepal Police, 5% in the Armed Police Force, and 3.2% in the Nepalese Army (Ministry of Peace and Reconstruction,1325 ACO Group Saathi, 2014).

These figures indicate that despite political efforts, Nepalese society is still struggling to realize gender mainstreaming, women's political participation and leadership. The Gender Equality and Social Inclusion Report by the United Nations and the Nepalese Ministry of Women, Children and Social Welfare (UN Women et al., 2017) expresses that rigid social norms, resource constraints, inadequate support systems, and lacking mentoring support are hindrances for anchoring GM into socio-political practice.

This gap is visible in education and academia as well. According to the Nepal Living Standards Survey, the overall literacy rates of males and females are 71.6% and 44.5%, respectively (National Planning Commission, 2011). Women represent 16% of undergraduates and achieve11% of doctorate degrees in engineering sciences, less than 22% of doctorate degrees in mathematics and physical sciences, and 28% of undergraduates and 15% of doctorate degrees in computer and information sciences (Prasai, 2019).

7.2.2 Norway

Norway, as well as Nepal, GM advocates "governance feminism" (Prügl 2011). In the past decades, GM and GE concepts have entered state- and state-like institutions, making gender an issue of structural, organizational, and governmental concern. Within this framework, Norway, contrasting to Nepal, can, for example, legally sanction public institutions that do not develop GE action plans and report their status regularly, which results among others in comprehensive GE policies, strategies, and action schemes in academia. Most Norwegian universities established their first GE plan in the mid-1970s (Nielsen, 2017).Besides regulations, the Norwegian government is providing economic incentives for universities hiring women in science and technology (S&T) disciplines. Correspondingly, NTNU, a technology-focused university, is offering start packages and scholarships for female researchers in S&T positions, e.g., to aid associate professors to become full professors. NTNU also offers a career mentoring program for women in (permanent) associate professor positions, yet there exist no programs for (temporarily) employed female PhDs and post-docs. Further, the institution established action plans to promote gender equality and address sexual harassment. In 2018, NTNU researchers from the Department of Interdisciplinary Studies of Culture (KULT) published the report "Gender balance from below – a toolbox" (in Norwegian only), (Øyslebø et al.), and some NTNU departments have adopted specific strategies suggested in this report. KULT is also conducting Master's courses on Gender and Culture.

Despite these efforts, female researchers' rates remain low in S&T disciplines in Norway as the report "Gender balance among researchers in Norwegian academia" indicates (Steine et al. 2020). For NTNU, the picture is similar; in 2019, only 27% of the full professors and associate professors were female (Suboticki et al. 2021).

7.3 Gender mainstreaming in sustainable planning and design

A renewable energy transformation has the potential to be a source of sustainable and equitable wealth creation for Nepal while diminishing climate change and reducing the country's need to import fuels from abroad. However, socially unequal access to energy solutions can avert these benefits. The lack of clean energy follows geographic, ethnic, class, and gender-based divisions and is often palpable in poor and rural areas, with corresponding implications for health, inequity, education, and mobility. Further, national policies and programs in Nepal continue to treat energy policy as gender-neutral in terms of impacts, failing to recognize the differing roles and needs of women and men in energy management. In summary, policies and programs for social inclusion and gender mainstreaming exist but are often not sufficient to bring about transformative changes (Mahat, 2003). Analyzing different needs, capacities, and use of energy sources and services as means for inclusion and identifying barriers that prevent women and excluded groups from participating in planning, design, and maintenance of energy solutions was hence a main part of the MSESSD program.

Tribhuvan University (TU) is the oldest and most renowned University in Nepal. It was established in 1959 and started affiliating with public colleges in 1971. Almost 87% of the student enrollments are in Education, Management and Humanities, while enrolment in Science and Technology, Engineering and Medicine are increasing. Against this backdrop, and with the motivation to mitigate the policy-practice gap at IOE, the Master's and PhDprogram "Energy for Sustainable Social Development" (MSESSD) was established in 2015. The funding from NORAD (Norwegian Agency for Aid and Development) ended in 2020; however, the Master's program is continuing as part of the regular IOE curriculum. So far, 60 Master's students from architecture and engineering (60% female students) and 8 PhD students (two female candidates) have graduated.

The goal of MSESSD is to provide Nepalese society with qualified professionals, who facilitate the design of clean energy solutions and contribute to good energy governance adapted to national, regional, and local stakeholder requirements. The project integrates several activities that encourage women and men to work with gender mainstreaming. The following activities are at the core of MSESSD:

1. Creating academic opportunities for female students and researchers.
2. Updating administrative personnel to effectively include gender mainstreaming.
3. Sensitizing female and male students for gender issues in energy systems.
4. Integrating female and male stakeholders in the design process.

The rate of female students who completed an MSc program at IOE, Tribhuvan University between 1998 and January 2015 was 10% out of 180 students in total. In order to enhance the overall inclusion of women in the institution, the program, and collaboration with local stakeholders, MSESSD pursued the following strategies:

1. For recruiting for the Master's program, female bachelor students were contacted directly before the program started and encouraged to apply, among others by the program coordinators. The students and their families were informed about conditions and possibilities (e.g., scholarships) as well as of existing and planned gender initiatives at IOE. Precisely, 50% of Master's scholarships were earmarked for female students. Further, a part-time gender advisor was appointed for female students, PhDs, and post-docs giving them career advice and support for day-to-day management in academia.
2. On an administrative level, NTNU hosted two courses on project management, targeted at administrative personnel at IOE. The two-week courses in 2016 and June 2018 comprised 50% female participants from eight participants in total. They include gender-mainstreaming information and strategies for academic education provided by NTNU's Gender Equality Adviser.
3. To sensitize the students to gender issues in energy systems, the project thematized gender mainstreaming in a specific course "Stakeholder Inclusion: Gender Mainstreaming and Energy Management" and integrated GM in several of the case studies in the Master's theses. MSESSD wanted the students to design end-user-friendly renewable energy solutions that can benefit local stakeholders and contribute to generate and enhance their livelihoods. For this reason, case studies that focused on smaller-scale technologies, which female stakeholders can take control of and manage themselves to meet their needs were favored.
4. The stakeholder collaboration and inclusion approach at Master's and PhD levels and for theses comprised collaboration with local women's groups such

as, e.g., Ama Samuah (Mothers group) in planning, decision-making, and management processes. The project met women's needs not just practically by providing better technology, but also strategically by providing solutions that enhance their social and economic status and their decision-making authority. Women in Nepal are the main managers of energy-related household chores and local businesses and are thus experts of (sustainable) practices within their specific environments (Keitsch, Singh, 2014) and possess great potential as drivers for clean energy.

7.4 Results and reflections: "Stakeholder inclusion: Gender mainstreaming and energy management (EL3)"

The course is one of the four elective courses in the MSESSD Master's program. From 2015 to 2021, the course was chosen by approximately 75% to 90% of the students (average class size 20).

The aims of EL3 are to convey basic GM theories; introduce novel practice-based methods for gender mainstreaming; discuss feasible strategies for stakeholder engagement; and encourage reflection on practical outcomes and the effectiveness of proposed solutions.

Besides GM theory, EL3 emphasizes the methodologies for collaboration with marginalized and female stakeholders from ideation to solution phase, partly guided by a "village perspective" (Hertel et al., 2019). This view assumes that engaging vulnerable population groups through community-level actors, such as non-governmental organizations, community-based organizations, and civil society organizations can be a feasible onset for collaboration with these actors. Strategies to include female stakeholders in the planning, design, and decision-making process and discuss energy innovation concerning local sustainable entrepreneurship were highlighted.

In terms of theory, the course comprised three parts: power issues, stereotypes about gender, and GM theories in and for the south. Power issues were discussed concerning Foucault's philosophy. Even though Foucault's "gender blindness" is criticized by some authors (McNay, 1992; King, 2004), his documentations of relations between discourse, power, and knowledge at a variety of levels also allow the students, to a certain degree, to analyze and deconstruct "grand (gender, M.K.) narratives" (Lyotard, 1984) on gender such as "the good wife" and the "selfless mother": "The grand narrative has lost its credibility, regardless of what mode of unification it uses, regardless of whether it is a speculative narrative or a narrative of emancipation" (Lyotard, 1984, Section 10, p. 37).

Discussing Foucault facilitates thereby telling alternative "narratives" on gender, and identifying generic and idiosyncratic aspects, rather than pointing out single sources of women's oppressions (Phelan 1990). For Foucault, "discourse" comprises written and spoken communication. It determines themes, directs understanding, and influences how ideas are put into practice. For example, a discourse on gender includes, among others, academic research, policymaking, novels, blogs, art writings but also "requisites" such as products and services for male and female consumers. All aspects of the discourse depend on the history and context as Foucault already discussed in his first book "History of Madness" (Foucault, 2009). Discourses are closely connected with power, who controls communication, who is admitted to the discourse, what contents are allowed, and how and by whom the *discourse is ritualized, visualized, manifested, and enacted*. Power also relates to the determination of what is the right and wrong standard, e.g., discrimination of those who are not "masculine" or "feminine" enough. In this setting, mainstream femininity is identified by some authors as a patriarchal set of discourses that women accept, negotiate, or resist (Leahy, 1994; Arya, 2020). Finally, Foucault refers to knowledge development. One condition for knowledge development is that societies have "regimes of truth" embodied in discourses. Regimes of truth follow rules such as giving certain types of discourses a special status. Further, they establish truth and falsehood criteria, assign rewards and sanctions for certain discourses, develop favored techniques for gaining truth, and assemble authorities to establish the truth.

The first analytic part of Foucault's theory in the course is supplemented with Foucault's view of the discourse-power-knowledge-constellation as exercised on different societal levels rather than being top-down domination. This is a productive and dynamic rather than a repressive and static perspective (Sawicki, 1988). The dynamic character of discourses *makes them negotiable for alternative visual and physical interpretations* (for example with help of design). This is important for both, moving toward a contextualized and flexible interpretation of GM and realizing the designer's, planner's, and architect's professional contribution and responsibility to the discourse-power-knowledge debate. Foucault shows several examples, how these professions contribute to shaping discourses (Foucault, 1984).

In EL3, the disciplinary confinement of the body is chosen to explain an ever-increasing and intense power through discipline. Design, planning, and architecture contribute among others to the development of drill mechanisms to manipulate, train, and shape the individuals' behavior by introducing precise regulations for how to walk, eat, write, speak, and work. And, as Foucault emphasizes, disciplinary techniques used to "discipline" prisoners are not essentially different from ways to train and shape scholars, students, soldiers, and workers.

One task of the participants in the course was to discuss the project "Proposed Whole Life Timers Prison in Dungeness" (Prince, 2011) with regards to Foucault's ideas on discourse, power, and knowledge. The project is a final Master'sproject in architecture and illustrates an example of both disciplinary measures regarding the mechanisms above and the role of the architect. The proposed prison building is situated in Kent, UK, between the National Grid's 400kV transmission lines that come from the Nuclear power station and shall house 45 life-sentenced prisoners. The clip accompanying the project is called "End of the line" and Prince explains:

> This project responds to a sensitive brief and is in no way a proposal for a new method of justice and detention. It is an architectural response to a varied context (the site) and factual information about the proposed main

> users ... The building seeks to combine the alien and surreal nature of the site with the housing of the most notorious offenders that exist in the United Kingdom today. The method as to how they are detained and disciplined is to this date constantly argued.
>
> *Prince, 2011*

The first part of the theory approach concluded with Foucault's concept of "biopolitics," i.e., power to control the life of a specific population (Foucault, 2008). Biopolitics is the exertion of state control over the functions and processes of life. Control over women's reproductive rights (abortion, birth control, etc.) is an example of biopolitics, and the force exerted by biopolitics over women's bodies is biopower. Biopower is exercised on all levels and in more or less subtle ways. While macro levels discipline via laws, micro levels make individuals obedient through social norms, -expectations and -stereotypes.

The second part of the theory approach of EL3 considered masculine and feminine stereotypes7. The goal was to make the students aware in their fieldwork, how stereotypes are represented in contextual settings. Stereotypes are open or hidden characteristics and interpretations and beliefs about groups of people because of their race, condition, nationality, body, etc. An advantage of a stereotype is that one can respond rapidly to situations because of similar former experiences. A disadvantage is that stereotypes ignore differences between individuals, i.e., by making generalizations and that negative stereotypes can lead to stigma. Stereotypes in GM discourses can influence people's behaviors, impact their subjective experiences of gender, and legitimate interventions, thereby moving back into the realm of biopower and -politics.

The third part of the theory approach was based on the debate around the efficiency vs. the equity-led approach. The efficiency approach focuses on actual needs and suggests technology alterations toward the pro-environment and the pro-poor. The goal is to identify needs and then provide solutions, which women can use, and control. It is the job of the planner, not the beneficiaries to find a solution that benefits both men and women in some way, often termed as "duality." This approach is characterized as conservative and static, rather uncritically accepting a societal status quo. It is referring to gender fixed roles and mainstream economy/technology-oriented (Chant et.al 2012). The equity-led approach aims to meet strategic needs rather than meeting specific problems and attempts to initiate meaningful participation and integration of women in decision-making processes. Interventions are directed to increase women's productive time. This approach promotes positive discrimination of female stakeholders in, e.g., resource allocation, incorporation of gender strategic interests in local policy documents. It can be characterized as progressive and dynamic, and critically challenges existing constructs and dominant ideologies. Alternative economy/technology is favored (Mahat, 2003).

Within this theoretical framework, EL3 introduced methods such as Mobile ethnography, Scenario workshops, User testing, and Cultural probes well as presented main policies and practices in the renewable energy sector in Nepal in both urban and rural contextualized settings. The course assignment, done group-wise, comprised the following tasks: to reflect on GM theory, to conduct a field study, to write a report on applying new stakeholder inclusion methods in a case study area in Kathmandu valley, and to present and reflect on findings and experiences for gender mainstreaming and energy-topics. All tasks had to sustainability topics and their links to GM.

7.4.1 Students' work and reflections

The following section presents some results from the groups' research, fieldwork, and reflections from the Spring Semester 2016. The three-day fieldwork took place in the traditional Newari settlement of Kirtipur, Kathmandu Valley, and comprised among others the application of mobile ethnography as a novel research strategy. Information was also gathered on basis of observation, interactions, informal interviews with local stakeholders (here residents and local groups), photographs, audio, and video recording. The goal was to understand GM in peri-urban sectors related to topics such as energy and water consumption, waste management, and use of public space.

The student reports started with literature reviews on GM concepts and policies, internationally and within Nepal. Despite multiple ethnicities, gender roles in Nepal mostly follow the Hindu caste system, consisting of Brahmins (scholars and priests) at the top, Chhetris (warriors), Vaishyas (merchants and traders), and Sudras (peasants/laborers). Everyone outside this system is regarded as impure. This group is referred to as "Untouchables," they call themselves "Dalits" (Hindi: "oppressed or broken").

The students described that despite the abolition of caste-based discrimination in Nepal in 1963, social hierarchies and divisions prevail, vesting power and privilege mainly to Hindu and Newari ethnicities. The patriarchal caste system also assigns women a subordinate status and position in society, which is ensuing in state-based discrimination. Two student groups researched social exclusion specifically related to energy and water consumption and indicated that exclusion in this sector is mostly concerning women and poor people.

> Women are responsible for fetching fuel and water for their households as well as for engaging in various types of microenterprises. Since the burden of providing energy to fulfill their household needs fall disproportionately on women, who spend a significant amount of their time and effort in collecting fuel, the opportunity cost is high for them as it takes them away from employment, education, and other activities for "self-improvement." Further, continued dependence on traditional biomass has detrimental effects on women's health through indoor air pollution caused by smoke and unhealthy workplaces. Yet, socially constructed perceptions that consider

> modern energy as "men's domain" limit the opportunities for women to take full advantage of new energy sources, particularly in entrepreneurship.
>
> *Tharu et al., 2018, p. 12*

The literature reviews were followed by the fieldwork in Kirtipur to investigate if and how findings from theory were noticeable in a real-life setting. Kirtipur's name derives from Sanskrit Kirti (glory) and pur (city). The city is located 5km southwest of Kathmandu and has a total population of 67,171 (Nepal Population and Housing Census, 2011). Kirtipur is inhabited and dominated by the ethnic group of Newars. Newars are earliest settlers in the Kathmandu Valley, and Kirtipur is center of Newari culture with its own language, traditions, and values. The city has many public meeting places, and informal gatherings of men's and women's groups during the evening and the holidays are frequent.

Gender roles are a crucial part of the Newari culture. Women are considered very important for managing household tasks to community development tasks such as conducting awareness camps, health camps, etc. Women are also the decision-makers in most families and at a lower community level. In the past, they were contributing through different organizations such as Ama Samuha (mothers group), MahilaSamuha (women's group), Buhari Samuha (daughter-in-law group), etc. Yet, with an increasing number of women entering economic professional life in Kathmandu since the late 1990s, traditional groups, tasks, and gender roles start transforming in the study area and men and women have no cultural templates regarding their new social roles in the study area. This generates tensions but also opens up new forms of interactions between male and female stakeholders.

The student groups applied interviews, observations, and mobile ethnography and analyzed the energy sector in terms of gender roles, stereotypes, best practices, and opportunities and challenges. In Kirtipur, women are the head of the household activities and the primary energy users, who have to deal with the efficient domestic use of fuels. Women have thus expertise in fuel-saving techniques, while men are responsible for buying the energy fuels and are active in delivering them from the source to the house. Liquid petrol gas (LPG) in containers is predominantly used as fuel for cooking purposes, which is the case for Nepal in general. Wood is not used frequently nowadays in Kirtipur, despite festivals and mass cooking. Firewood is also used by elderly women for the preparation of Newari cuisines such as Bara and Alcohol because of the assumption that it improves the taste. Ashes from the burnt fuelwood are used as compost in the kitchen garden, which can be considered as good practice. In Kirtipur, Grid Electricity is the main source, many households also have solar PV cells and solar water heating systems. Electrical appliances are operated by women, such as pumping water by machines, electric lights, washing machines (those who have), and other daily household electrical appliances, only minor maintenances are done by women. Energy awareness is more prevalent among female stakeholders in Kirtipur. Both women and men are aware of more thermal comfort in traditional buildings than in modern houses.

Energy use also plays a vital role in cultural and religious activities in Kirtipur. One of the cultural and religious activities is the cremation practice of the Hindu religion that has been practiced in the city for centuries. The burning of the body with firewood, straw, and oils at the Ghats is normally done by men, illustrating the stereotype that women carry out domestic activities, while men perform public rituals. Another stereotype that the groups pointed out relates to women and work. Economically, LPG is used in most of the women's SME entrepreneurship activities, where needed. Female stakeholders are supposed to be local entrepreneurs, yet rather working from home, than being employed in Kathmandu City, which is considered a task for males.

Opportunities for GM in Kirtipur relate to women's active in development programs and participation in different groups for implementing best practices such as:

- Reuse of fuelwood residue(ashes) in agriculture
- Reuse of construction materials in new construction
- Waste to compost practice
- Use of renewable energy such as solar and biomass
- Effort in making the traditional settlement area intact in the core city
- Starting of homestays in traditional style
- Community Newari Cuisine Restaurant – LAHANA and Live open museum

The involvement of local stakeholders, especially male and female youths in programs, is however considered as a challenge in the study area since young people tend to work and study in Kathmandu and Kirtipur is increasingly becoming a bedroom community.

Relating to their experiences and findings, the students conclude that being a traditional settlement, Kirtipur has implemented some good practices enhancing gender mainstreaming. Provision of women-friendly space (i.e., a small building constructed especially for women) for meetings, and training in handicrafts and knitting, is mentioned as possibilities for GM.

In the author's opinion, this also shows that the Nepali students often reflect on capacity building for women in more conservative ways. The students were also aware that a lot more can be done such as training women in small-scale industries, for example, milling and oil production to bridge the gap from unskilled to skilled worker. Further, programs on gender mainstreaming, increasing the women's literacy rate, and property rights for women were suggested.

Regarding the new methods applied, it contributed for some students to unexpected insights into one's own role. As one female student wrote:

> The methodology enabled me to explore the surroundings and the people but more than that I learned about myself .... going into the groups of men and inquiring about their activities was something that we avoided. ... when I went to interview the driver and he and his friends were busy in the gossip,

> I didn't feel comfortable just going and talking, I rather preferred waiting for the driver outside and all these situations occurred just because I am a female and I identified with it. The even more surprising or rather shocking thing is that I didn't realize it when I was there nor I realized this gender consciousness stuff when I recalled the situations only when I read the literature to prepare this report I came to know about. Thus even the 'so-called literate people like me' who have been talking about gender mainstreaming all the week are being affected by gender disparity and not being conscious about it, so I couldn't help wondering about the conditions of the women in rural areas.
>
> *Uprety, 2016*

7.5 Summary from integrating GM strategies in design and architecture curriculum and academia and Nepal

In terms of didactics, the above statements indicate that integrating GM in design and architecture curriculum needs increased awareness and further discussions of (so far tacit) gender stereotypes. Applying novel methods such as visualization techniques and mobile ethnography can facilitate discussions with both students and local stakeholders. Mobile ethnography also seems to make students and researchers more aware of the fact that even if the other person thinks differently, one can still cooperate via the common interest in finding solutions, shared problem space and choices of tools, and through conducting common activities within the given operational setting. Methodologically, the mobile ethnography approach can contribute a transdisciplinary understanding of the architect's and designer's professional role, which in turn might advance collaborative and inclusive GM practice. The students and teachers have not conducted research or interviews with the stakeholders about if and how mobile ethnography contributes to transdisciplinary in GM collaboration (Watts-Englert et al., 2014); this remains a topic for future research and stakeholder inclusion projects in planning and design.

In summary, comments from the IOE Master's students encourage future activities to integrate GM in S&T education and fieldwork. In engineering, design, and architecture, GM will often relate to solution-oriented approaches, following an efficiency- rather than an equity-led approach. The latter can, however, guide case studies on the PhD level, including discussions of stereotypes, gender, and GM strategies for structure and systems improvement, which has been done in a recent MSESSD PhD dissertation (Shrestha et al. 2020).

Practically, the feasibility of NTNU GM approaches in IOE/MSESSD contexts is highly dependent on the level of decision-making. Including more women in research and education worked well in the program on a basic level as "Gender balance from below." On the middle management level, the project contributed to developing a Gender Action Plan at IOE. Yet, IOE is part of Tribhuvan University and TU has apparently no GM/GE action plan. Missing GM coherence in education institutions reflects also a weak link between academia and policymakers in Nepal.

On the plus side, the explicit collaboration with societal stakeholders on GM topics in MSESSD created increased GM attention and awareness at IOE. This concerns administration, teachers, researchers, and stakeholders as well as architecture, engineering, and design students and resulted in GM sensitive design and planning solutions on the Master's level, and theory development and analyses on the PhD level.

Relating disciplinary education and research and GM in practical social settings is lacking at NTNU and so is the thematic integration of GM in the technology disciplines. NTNU still pursues a thematic and disciplinary silo approach, where GM in education and research takes place in the humanities only.

Conclusively, awareness of power structures is necessary to avoid imposing "western" gender ideologies when transferring GM strategies from Norway to Nepal. In the MSESSD program, this was especially true for GM relating to stakeholder collaboration. Not only are western GM values and theories often far from the experienced reality of the Nepalese stakeholders but gender inequalities can be hardly addressed without profound contextual knowledge. Yet, becoming culturally acquainted is difficult in a country with 101 ethnic groups speaking over 92 languages. Even IOE students and teachers come from different cultural backgrounds with unlike gender values and challenges. For the MSESSD program, two maxims to prevent competence imperialism were: to include the whole community, especially local women and women's groups as allies of change in the very beginning of every project and fieldwork, and to provide solutions that benefitted the whole community (Keitsch, 2021).

In academia, the global value of the "The 2030 Agenda for Sustainable Development," which was adopted by the 193 Member States of the United Nations, including 17 SDGs (United Nations Department of Public Information, 2015) provided a common onset to mitigate ideological biases and supremacies. SDG 5 "Gender Equality" worked here as a platform for IOE and NTNU to integrate GM in architecture and design curricula with the ambition to advance students' critical thinking capacity, apply new methods to detect gender stereotypes, and closely work with local stakeholders to design solutions, which increase GM in contextualized everyday sustainability settings.

Acknowledgments

I am grateful to Sangeeta Singh for her constant help and support in implementing GM at IOE. Further, I would like to thank PhD Bind Bindu Shresta for valuable comments, the Ama Samuah group in Kirtipur, Nepal, and all IOE MSESSD Master's students from the batches 2016–2018.

References

Arya, S. (2020), Dalit or Brahmanical Patriarchy? Rethinking Indian Feminism, CASTE, *A Global Journal on Social Exclusion*, Vol.1, No. 1, pp. 217–228.

Bhattachan, K. B. (2005), Social and Economic Conditions of Dalit Women, in: Pramodh M.Devkota (ed.) *Dalits of Nepal Issue & Challenges*. Kathmandu: Feminist Dalit Organization, pp. 65–89.

Bjørvik, O. (2017), The Role and Effectiveness of National Action Plans as Soft Law Tools to Implementing United Nations Security Council Resolution 1325 on women, peace and security – With a case study of Nepal's National Action Plan on Implementation of UNSCRs 1325 and 1820, Master's thesis, accessed 4 May 2021, www.duo.uio.no/bitstream/handle/10852/56517/1/537.pdf

ChantS., and Sweetman, C. (November 2012), Fixing women or fixing the world? 'Smart economics', efficiency approaches, and gender equality in development, Gender and Development, *Beyond Gender Mainstreaming*, Vol. 20, No. 3, pp. 517–529.

Council of Europe. (2021), Gender Equality and Gender Mainstreaming, accessed 20 July 2021, www.coe.int/en/web/gender-matters/gender-equality-and-gender-mainstreaming

Foucault, M. (1984), Of Other Spaces: Utopias and Heterotopias, *Architecture /Mouvement/ Continuité*, accessed 4 May 2021, http://web.mit.edu/allanmc/www/foucault1.pdf

Foucault, M. (2008), *The Birth of Biopolitics, Lectures at the Collège de France, 1978–79*, English translation published in 2008 by Palgrave Macmillan, New York.

Foucault,M. (2009), *History of Madness*, translated by Khalfa J., Routledge, New York.

Hertel, C., Bacq, S., and Belz, FM (2019), It Takes a Village to Sustain a Village: A Social Identity Perspective on Successful Community-Based Enterprise Creation, Academy of Management Discoveries, Vol. 5, No. 4, published online https://doi.org/10.5465/amd.2018.0153

Keitsch, M., and Singh, B. (2014), Sustainable Energy Design, Stakeholder Collaboration, and Gender Mainstreaming in Nepal. EnGenderingCities; Designing Sustainable Urban Environments, Rome 25–26 September 2014, International Conference organized by genderSTE and the Italian Presidency of the European Union.

Keitsch, M. (2021), Transdisciplinary Collaboration and Ethics, in: Keitsch, M., Vermeulen, W. (eds). *Transdisciplinarity for Sustainability*. Abingdon: Routledge, pp 53–71.

King, A. (March 2004), The Prisoner of Gender: Foucault and the Disciplining of the Female Body, *Journal of International Women's Studies*, Vol. 5, No. 2, pp 29–39.

Leahy, T. (1994), Taking Up a Position: Discourses of Femininity and Adolescence in the Context of Man/Girl Relationships, *Gender and Society*, Vol. 8, No. 1, pp. 48–72.

Lyotard, J.F. (1984), *The Postmodern Condition: A Report on Knowledge*. University of Minnesota Press, Minneapolis.

McNay, L. (1992), Foucault and Feminism: Power, Gender and the Self Polity Press, in: Cambridge Ramazanoglu, C. (ed.) (1993) *Up Against Foucault: Explorations of Some Tensions Between Foucault and Feminism*. Routledge, London.

Mahat, I. (2003), Women's Development in Nepal: The Myth of Empowerment, *The Fletcher Journal of International Development*, VOLUME 18 – 2003, pp. 67–71.

Ministry of Peace and Reconstruction and Action Group Saathi (2014), Nepal National Action plan (2014), accessed 20 April 2021, https://saathi.org.np/wp-content/uploads/2018/10/saathimidtermreportenglish.pdf

National Planning Commission, Central Bureau of Statistics (2011), Nepal Living Standards Survey, accessed 20 April 2021, https://cbs.gov.np/nepal-living-standard-survey-2010-11/

Nepal Law Society, International IDEA, and UNDP (2021), Nepal's Constitution of 2015, accessed 4 May 2021, www.constituteproject.org/constitution/Nepal_2015.pdf

Nepal Population and Housing Census (2011), accessed 4 May 2021, https://unstats.un.org/unsd/demographic/sources/census/wphc/Nepal/Nepal-Census-2011-Vol1.pdf

Nielsen, M. W. (2017),Scandinavian Approaches to Gender Equality in Academia: A Comparative Study, *Scandinavian Journal of Educational Research*, Vol. 61, No. 3, pp. 295–318.

Phelan, S. (May 1990), Foucault and Feminism, *American Journal of Political Science*, Vol. 34, No. 2, pp. 421–440.

Prasai,Y. (2019), Education for women, Himalayan Times, 8 November 2019, accessed 4 May 2021, https://thehimalayantimes.com/opinion/education-for-women

Prince, B. (2011), "Proposed Whole Life Timers Prison in Dungeness, Kent, England" Master's thesis, www.youtube.com/watch?v=cPA8JUCDbTM

Prügl, E. (2011), Diversity Management and Gender Mainstreaming as Technologies of Government, *Politics and Gender*, Vol. 7, 71–89.

Sawicki, J. (1988), Feminism and the Power of Discourse, in:Arac,J. (ed.) *After Foucault: Humanistic Knowledge, Postmodern Challenges*. Rutgers University Press, New Brunswick and London, pp. 161–178.

Shrestha, B., Bajracharya, S., Keitsch, M., and TiwariS. (September/October 2020), Gender Differences in Household Energy Decision-Making and Impacts in Energy Saving to Achieve Sustainability: A Case of Kathmandu, Sustainable Development, Vol. 28, No. 5, pp. 1049–1062.

SteineSæbø, F., Gunnes, H., and Wendt, K. (2020), Gender balance among researchers in Norwegian academia, The Nordic Institute for Studies in Innovation, Research and Education, Report Insight 20-2020, accessed 20 July 2021, https://nifu.brage.unit.no/nifu-xmlui/handle/11250/2722116

Suboticki, I., Lagesen, V., and Unander, T. (2021), En kartlegging og evaluering av tre kjønnsbalansetiltak ved NTNU, Rapport på oppdrag fra Utvalget for likestilling og mangfold ved NTNU. Trondheim 2021.

Tharu, B., Rajkarnikar S., Nakarmi, N., and Bhandari, S. (2018), Gender Roles in Energy in the traditional settlement of Kirtipur, MSESSD EL3, Project Report May 2018.

The Electoral Knowledge Network (2021), Nepal Constituent Assembly Election 2008, accessed 30 March 2021, https://aceproject.org/ero-en/regions/asia/NP/nepal-final-report-constituent-assembly-elections

United Nations Development Program, UNDP (2021), Nepal SDG 5, accessed 20 April 2021,www.np.undp.org/content/nepal/en/home/sustainable-development-goals/goal-5-gender-equality.html

United Nations Department of Public Information (2015), Historic New Sustainable Development Agenda Unanimously Adopted by 193 UN Members. Press Release. New York, 29 September 2015, accessed 20 July 2021, https://sustainabledevelopment.un.org/content/documents/8371Sustainable%20Development%20Summit_final.pdf

United Nations Climate Change (2015), Paris Agreement, accessed 20 April 2021, https://unfccc.int/process-and-meetings/the-paris-agreement/the-paris-agreement

United Nations Women (1995), The Beijing Declaration and Platform for Action (1995), accessed 4 May 2021, https://beijing20.unwomen.org/en/about

United Nations Women (1995), The Beijing Declaration and Platform for Action (1995), Section 16, accessed 4 May 2021, https://beijing20.unwomen.org/en/about

United Nations Women, Nepalese Ministry of Women, Children and Social Welfare (2017) Gender Equality and Social Inclusion (GESI) Profile For Humanitarian Action and Disaster Risk Reduction, accessed 4 May 2021, https://reliefweb.int/sites/reliefweb.int/files/resources/Nepal%20GESI%20Profile.pdf

Uprety, S. (2016), MSESSD, EL3 Reflection Report.

Watts-Englert, J., Szymanski, M., Wall, P., and Sprague, M.A. (2014), Collaboration for Impact: Involving Stakeholders in Ethnographic Research, *International Journal of Business Anthropology*, Vol. 5, No. 1, pp. 55–70.

Øyslebø, S., Sørensen, K., Lagesen, V., and Kristensen, G. (2018), Verktøykasse for bedre kjønnsbalanse, Project Report Trondheim 2018, accessed 4 July 2021, www.ntnu.edu/documents/1267482954/1279211967/Verkt%C3%B8ykassen.pdf/e8e4ea5e-8cb0-4045-a477-27b5d6aadc34

8

COMMUNICATION DESIGN TO FOSTER GENDER EQUALITY

Research and experimentation in the educational field

Valeria Bucchetti and Francesca Casnati

8.1 Introduction

The urgency represented by the production of media images which fuel discriminations against women contributes to what is defined as *gender power assimmetry* (Mulvey, 1975), a gap which in different ways keeps on wounding our current society and which, today more than ever, needs to be addressed as a priority. The visual messages conveyed by the media all too often fit into the lower part of the *iceberg of gender-based violence* through those manifestations defined as subtle and invisible, which include sexist advertising and, more generally, sexist language. These are manifestations that we have to deal with everyday and that we too often perceive as *normal*.

Those manifestations grow within the current socio-cultural framework, particularly in Italy, which lives a level of backwardness well documented by the data reported periodically by the "Global Gender Gap Report."

Despite the issue of gender discrimination has been at the center of the international debate for decades now (UN Agenda 2030 for sustainable development, Goal 5; Beijing Platform for Action, 1995; The Istanbul Convention, 2011; Beijing+ 25, 2020), the processes of growth and social inclusion are still slow, weighed down by a strongly male-oriented socio-cultural heritage, which is reflected and fed by the flow of media messages characterizing the environment in which we grow up, live, and learn to relate to each other.

The society of the third millennium is in too many ways still characterized by what Bourdieu called *male domination*.

The heterosexual and able-bodied white man (Criado-Perez, 2019) has a central role; this androcentrism is inherited from the cultural tradition that brings with it behavioral patterns, *habitus* (Bourdieu, 1998), and harmful stereotypes that represent a brake in relation to the inclusion. In 2020 in particular, we have

DOI: 10.4324/9781003174257-10

witnessed in various contexts a real downgrading in the recognition of women's rights. According to UN Women, the emergency situation dictated by the Covid-19 pandemic, and the lockdown imposed to address it, has globally encouraged the growth of violence against women, particularly forms of domestic violence.

In the institutional and political spheres, we are also witnessing increasingly frequent episodes of backsliding from the point of view of gender-based social discriminations. One of the recent cases concerns the position taken by the Turkey's President Erdogan, who in March 2021 overcame the Istanbul Convention, despite data on feminicides in Turkish territory still show a critical situation: 78 feminicides since the beginning of 2021 and more than 300 in 2020.[1] After this episode, other nations have expressed their desire to leave the Convention, because it conflicts with their values.

But even the wave of violence and sexism left in the United States by Trump's presidency, or minor issues such as the debate that has arisen in Italy around the *Zan* draft law[2] against violence based on gender, sexual orientation, gender identity, and disability, which has seen parties and exponents of the right take sides against it, demonstrating how the worldview linked to the patriarchal cultural tradition is still rooted, denying pluralism. In this hostile and distrustful mood when referring to gender equality, media images constitute forms of non-material domination, symbolic violence, which permeate the socio-cultural dimension and presuppose a social relationship in which the dominant exerts forms of indirect violence on the dominated, acted through the imposition of a certain worldview, social roles, cognitive categories, and mental structures.

8.2 A polluted media-sphere

In this framework, the main issue that as communication designers concerns us is the socio-cultural responsibility of the media images which pervade the socio-sphere, orienting the choices and opinions of the individuals.

As Nicholas Mirzoeff states in *How to See the World*, "the emerging global society is visual" and the images we produce as designers—but also as individuals who are part of a socio-cultural environment—are a vehicle for specific worldviews and have inevitable repercussions on the construction of individual and collective biographies.

The *communicative landscape* (Baule, 2012) that defines the everyday dimension is permeated by gender stereotypes which contribute to the definition of the Self (Camussi & Monacelli, 2010) conditioning behaviors, cognitive patterns, and the perception of Self and others within a social community.

At the international level, there are many appeals and measures that place the media—and the system of actors involved in the construction and distribution of media messages—at the center of a social responsibility system, emphasizing the urgency and need to eliminate all those sexist forms of representation which have serious repercussions on the role of women within society. It was with the "Beijing Platform for Action" of 1995 that interactions between women and the media were placed at the heart of the international debate. In the "Beijing Platform," in particular, the persistence of sexist stereotypes in the production of images and media messages was pointed out, and attention was drawn to the effects that violent images have on equality between men and women. In 1995, the Council of the European Union also expressed its concern about the representation of women in advertising, calling for the promotion of diversified images able to represent the complexity of reality. Followed by several European Parliament resolutions including those of October 2008 (2008/2038(INI)) and April 2018 (2017/2210(INI)) on the impact of marketing and advertising on gender equality which "highlights the importance of promoting media literacy [....] so as to encourage young people to develop critical thinking skills and to help them identify and denounce sexist representations and discrimination"; but also the 2013 resolution in which the European Parliament observed, despite appeals and measures taken, the absence of significant progress at the European and national institutional level regarding discriminatory messages conveyed by advertising. This backwardness was more recently underlined by the document produced in 2020 for *Beijing + 25*[3]:

> the European Parliament has highlighted persistent gender stereotypes in the media and called on the European Commission and Member States to take action. New forms of media also put women at risk of harassment, particularly those in public positions. The increasing prevalence of social media poses a challenge for regulators, as new rules need to be established to deal with the harassment that women suffer on these channels, including possible sanctions against media organisations.

In this international scenario, in itself still backward, the Italian reality is placed as the tail end. As Marilisa D'Amico highlights in "Una parità ambigua. Costituzione e diritti delle donne," despite other European countries are taking action to fight sexist communication at an institutional level, in Italy there is a serious lack of effective measures of protection, and the existing rules are too vague. In recent decades, we reached the threshold of communicative degradation, a saturation of polluting images that "finds structural reasons in a system which, over time, has seen technological transformations, concentration of power, reconversion of modes of use, the affirmation of models without an adequate elaboration on the level of non-mercified cultures" (Bucchetti, 2021; author's translation), models we tend to justify by appealing to the freedom of expression, stepping on the constitutional equality between men and women.

These stereotypical and degrading forms of representation generally respond to two different logical approaches.

On one hand, we observe a kind of conscious designing that aims, through the use of widely spread and accepted stereotypes, to activate a *winking* process between the sender and the receiver; on the other hand, however, there are images that unwittingly embody the designer's point of view, the models and canons that—as a subject belonging to a socio-cultural context—they has internalized and that he/

she unconsciously translates into the communicative project. It is precisely from this last point that the urgency of acting upstream of the phenomenon arises, the need to provide an education to designers in order to make them aware of their role of social responsibility. If it is actually true that the communicative project partially translates the designer's worldview, this cannot, however, justify the circulation of images that harm, even if unconsciously, gender equality.

Visual communication designers should be equipped with the tools for a critical reading of their own work, tools that enable them to design images that are sustainable and ecological within the communicative landscape.

From this point of view, communication design is able to provide a disciplinary contribution in order to foster the reconstruction of a fair and inclusive communicative landscape, acting on the training of new generations of designers—who will be responsible for re-shaping and decontaminating the media environment that characterizes and will characterize society—in order to lay the foundations for a new sensitivity that leads to the innovation of communicative languages, in a long-term approach.

In the past few years, the research group dcxcg[4] (Communication Design for Gender Cultures), which was born within the Design Department of the Politecnico di Milano, has been involved in developing a series of activities in the educational field which go in this direction, fostering a (self)critical and analytical approach to the project of visual communication, but also encouraging students—and therefore future designers—to work on projects of public utility, emphasizing the political dimension of communication design.

In this context, in 2014, a course in the Master of Science program in Design was introduced within the School of Design at the Politecnico di Milano, focusing on the relationship between Communication Design and Gender Cultures. It is the first course dealing with these issues in the field of design within Italian universities, and it provides a first proposal to prevent media pollution by responding to the resolution of the European Parliament of 2008 and subsequently to that of 2018, in which attention is focused on the need to raise conscious and responsible communication designers. The innovative side, as far as the Italian cultural context is concerned, lies in the specific kind of approach based on the relationship between two disciplinary fields. Gender Studies and Communication Design, interlacing, within the design culture, the sensibility for gender issues and the sensibility for the educational issue.

8.3 Stereotypes and communication designers: A vicious cycle

Before going deeper into some of the experiments developed in the educational field, it is useful to stress the dichotomy related to the role of stereotype in the area of visual communication. We live in an increasingly complex reality that undergoes dynamics of cognitive simplification allowing us to interpret reality, communicate, and be understood. The communication designer himself applies simplification mechanisms that are essential to make the message incisive and decodable by a wide range of recipients. Stereotypes are therefore part of our cognitive system; their construction originates from an individual and collective process of categorization and simplification of reality: through the lens of subjective experiences or intentions a quality, a role, a behavior, is associated with a gender.

The simplification of reality is therefore a cognitive act, essential to the communication designers who handles stereotypes on a daily basis in order to construct their own narratives, and if they do not have the proper cultural basis, they risk being an accomplice to the spreading of a deformed vision. The mechanisms of simplification take on a negative value when they become no longer functional to the effectiveness of the communication but fall into what Maldonado calls *hypersimplification* (Maldonado, 1959) of communicative languages that leads to the semantic banalization of the message.

Zingale in "Immobili visioni"[5] defines the stereotype as "an ideologically oriented connotation" which requires:

> (I) an arbitrary generalization, because a particular trait is elevated to a general value; (II) a social sharing, because it is assumed by one social/ethnic group against another group; (III) a semantic inflexibility, because when it imprints itself in the consciousness it is difficult to modify it.

The strength of the stereotype lies precisely in its fixity, and its meaning was sanctioned under the presumption of having fixed it once and for all, despite the fact that it does not fit well with the multiplicity of the real that it presumes to represent.

Another peculiarity of the stereotype is its pervasiveness, the insistence with which it is reiterated until it is "imprinted in the consciousness" and perceived as *normal*. A mechanism of internalization and addiction to the visual stereotype that Kanizsa defines as *new realism*, or the attitude of those who define as *natural* certain representative models. Representative repetitions thus end up influencing the perception we have of reality, entering a vicious circle that has repercussions on the way we relate within society.

In the case of gender stereotypes, the media provide a model of women based on an androcentric and sexist worldview, which from the earliest years of life becomes part of our visual universe and constitutes the lens, more or less thick, through which we observe the world. The designer himself unconsciously wears these lenses, fitting within a flow in which "new images are strictly defined in relation to those that have formed the sensibilities and mental archives of those who have generated them" (Bucchetti, 2021:46; author's translation). If we try to schematize the phenomenon, what best represents the dynamics according to which the stereotype develops and is strengthened is the symbol of ∞ (Bucchetti & Casnati, 2019). This representation allows us to imagine the stereotype within a continuous flow that progressively reinforces it. One half of the infinite represents the socio-cultural environment in which we grow up and from which we absorb cultural models and stereotypical forms, and the other half refers to the production of media messages, where we act as designers using stereotypes in the construction of messages, producing communicative artifacts that in turn fit into the socio-cultural environment fueling the stereotype.

Every time a stereotypical model transits on the intersection point between the two rings, it strengthens, it becomes a socially shared opinion, a *status quo*, influencing individual and collective biographies and reflecting on social identity. The scheme of this self-reinforcing flow allows us to visualize the critical points and to facilitate the deconstruction of the phenomenon and provides support in understanding the role of communication design within the process in order to identify the actions that lead to a slacking of the *vicious* circle and its reversal into a *virtuous* circle.

In this light, communication design in the educational field has the task of working upstream, spreading among the youngest a new design sensitivity, where awareness and self-reflection play a decisive role.

8.4 Breaking the endless cycle: Experiments and solutions from the educational field

Acting on the education of individual visions and awareness is fundamental and delicate. It means working in the ethical and political dimension of design—the ethics of images, the ethics of communication, the ethics of gender, the ethics of education—by training students in the direction of new, non-imposed views. It means providing adequate tools to independently develop culturally sustainable communication solutions.

According to Dewey's perspective, it is the responsibility of the educator to foster cultural changes that are accepted, internalized, and acted upon by the students, in our case, the designers. Education, at any level, is an indispensable component of ethical and social development. As Baule reminds us in the introductory chapter of "Rimediazioni gender-sensitive" (Caratti, 2015), the task of those involved in the education of communication designers is to

> guide on paths and not to get lost in destinations, to provide alphabets for writing and reading, including 'visual' ones, and not merely made up phrases or dogmatic constructs. The conformism inherent in the vision of the media and their stereotyped content cannot be contradicted by an educational conformism equipped with ready-made solutions, but by tools that can activate critical intelligence.

The challenge lies in guiding the students along a path of awareness and critical reflection on communication design and on their own work, in a perspective that combines professional and intellectual training. The mission is therefore not to provide notional and prescriptive information, but rather to work on the tools of design through a direct dialogue between teachers and students aimed at orienting and guiding the designer in a process of individual growth.

The course "Communication Design and Gender Cultures" has represented, and still represents, a fertile ground for the experimentation of educational activities that build on these principles, allowing us to make a continuous reflection on the strategies that university can put in place to prevent the circulation of further representative polluting models. The educational model is based on teaching methods that combine the theoretical and practical aspects of the discipline of design, working in the area of intersection between communication design and gender studies according to a multidisciplinary approach. In some respects, the structure of the learning model can be traced back to the Pyramid of Active Learning while remaining flexible and open to continuous small redirections. The course includes a theoretical part of ex-cathedra lectures aimed at laying the ground for a critical reading of media images and providing tools for the deconstruction of gender stereotypes. A second part of the course is rooted in the practice and theories of "learning by doing," during which emerges what Dewey calls "organic link between education and personal experience." Students are guided through a path that starts from actions of observation and critical analysis of the forms of representation typical of the socio-cultural context in which we live, and it is articulated in a series of steps that can lead to, on one hand, the experimentation with translational models and communicative languages that allow to bring to attention the forms of gender discriminations found during the first observation activity, and on the other hand, experimentations with methods of direct observation on the *making*, on the acting of the designers. Activities that aim to investigate the influence of unconscious gender bias on the communicative artifact, in order to foster constructive self-reflection.

8.4.1 Unmasking stereotypes to denounce sexism in the media

In the case of gender issues, the dimension of critical-reflection has been assumed as a real mode of cultural contestation against the dominant models: through the deconstruction of the forms of media representation, it is therefore possible to bring to the attention of the community all those stereotypical forms that reflect attitudes, roles, degrading social models. In this approach, communication design takes on the responsibility of examining the cultural models it spreads. From a pedagogical point of view, the designer is led to develop deconstruction tools and revision processes that can reveal the implicit, with positive effects on design thought itself. The intent is to guide the designer to intercept grammars, languages, and compositional rules that produce discriminating and misleading meanings. Through the deconstruction of the logic underlying the forms of stereotypical representation and their re-assemblage through narrative constructions, it is possible to make evident the short circuit of meaning in the mechanisms of the stereotype, going to weaken the flow represented by the symbol of infinity mentioned above.

When considering communication design in its dimension of a translational system (Anceschi, 1981; Baule & Caratti, 2017), young designers are led to experiment with grammars and communicative languages which, through the use of rhetorical devices, are able to denounce the trap of the sexist stereotype.

It is useful to mention as an example the work carried out by a group of students who observed the representation of women in periodical magazines on the subject of science and technology. In this specific case, the teamwork analyzed the Italian best-selling STEM magazines, from which they isolate the dominant representative models, identifying several levels of criticality which goes from the numerical inferiority of female figures compared to male ones, to the qualitative aspects concerning the few female figures present in the sample observed. Female figures who are precluded from any STEM-related activity and seem to have been included purely to please the male gaze. These are figures with hypersexualized traits and poses, who mainly play the roles of the object-carrying woman, the one of *ornament* to the article, the transfixed woman or, in the case where it relates to technology, the "unable" one. The focus fell specifically on the objectification of female bodies on three magazines focused on technological innovation. The students, by deconstructing stereotypical models and translating them across different semantic areas, identified a similarity with the visual languages—poses, expressions, camera frames, etc.—of the *soft-porn* area. This set the groundwork for building a narrative plot that could expose the unjustified hypersexualization of women in STEM fields.

The narration, or *counter-narration*, has been constructed by using a rhetorical mechanism that we can define *semantic flipping*. An interactive visual composition was designed which, through the use of a horizontal scroll, allows to place some female figures extrapolated from the covers of the STEM magazines analyzed above the "right cover." The trick that allows to unveil the stereotype resides in the choice of the covers: in the visual composition, covers of stem magazines and covers of soft-porn magazines were mixed together. It is impossible for the user to position the female figure in the correct way; each figure can be perfectly placed on both kinds of magazine, leading to disorientation. This rhetorical mechanism enables us to bring attention to the spread of misleading representational models by unmasking the stereotype. The students in this case carried out an exercise of deconstruction of the stereotype, isolating the recursiveness and experimenting with grammar and rhetorical keys in order to give life to a real counter-narration, aimed at denouncing the issue related to the objectification of women in STEM fields.

Other forms of reconstruction and denunciation have been developed over the years based on the same concept of deconstruction and reassembly, but which exploit different rhetorics according to the effect of meaning that the designers intend to produce. In any of these cases, these are activities that enable you to acquire, through the practice of design, tools that, if applied in the professional field, help in designing communicative artifacts with awareness, without falling into the stereotypes trick, generating further forms of harmful representation.

8.4.2 Direct observations to support the self-critical reflection: Designers and unconscious biases

Another key topic in the educational field concerns the ability and the inclination to the practice of self-reflection, a crucial issue in order to acquire awareness of what we are designing and of the effects that a communicative message will produce once circulating within the socio-cultural context. Self-reflection should distinguish each action of the communication designer, beyond gender issues. Hence, the desire, over the years, to introduce and experiment methods of direct observation of the *making* within the class, with the aim of investigating the practice of the designer himself, the ways in which we articulate the design thought and the unconscious prejudices that we inevitably bring with us and that are real traps in which even the new generations of designers are likely to fall if not trained in critical thinking.

In this case, the interpretative level concerns the *actors*, and it focuses on the observation of the community of design students in order to understand how they approach the project, from a perspective that considers attitudes, thoughts, dialogues, and unconscious gender bias. In short, these are actions aimed at critically observing the work of the designers themselves and their design thought, answering the generic question of "are gender stereotypes still being perpetrated by the younger generations of designers?" (Bucchetti, 2021).

Students are encouraged to investigate the weight of common sense in the younger generations, of which they are a part, by observing the influence produced by implicit knowledge and the component of unconscious bias that is expressed during the development of design thought. Observing the actions of their own colleagues and peers allows the student to practice self-reflection, which is developed through the planning of direct observation actions and the subsequent analysis and deconstruction of these moments of observation focusing on the influence that unconscious bias has on design thought.

One of the activities that best exemplifies this type of experimentation focuses on the representation of the family, a particularly awkward issue in Italy, where, especially in recent years, there has been a violent debate about the concept of family and what is defined as "traditional family." The objective was to verify, following an observation of commercials transmitted by the Italian television, whether young designers placed in the situation of having to design an advertising spot fell into the same stereotype of the family that characterize advertising communication in Italy, namely the family composed of three/four members of which father–man, mother–woman and children. Generally, according to the widely diffused model, this ideal family nucleus also corresponds to very precise social roles and is referred to, in the local slang, as the *Mulino Bianco family*[6] since the food brand for years has based its communication on this kind of stereotype. A preliminary research phase was useful in collecting and documenting precisely these reiterated patterns and laid the groundwork for the subsequent participatory activity.

From an educational point of view, preliminary observation allows the students to apply tools belonging to design research, to gender issues and in particular allows them to develop a critical point of view on design artifacts themselves. But we want to bring attention to the direct observation activity developed by the students following the first research work. The point of interest and object of observation was the design thought enacted by other students from the same Communication Design degree program. Therefore, a group of students was involved, to whom the researchers submitted the script of a hypothetical commercial, without giving them any anticipation regarding the theme being investigated. The script was designed to be devoid of any gender markers or indicators that could guide its interpretation. The script described a typical situation found in food commercials, the act of eating breakfast, organizing the day and some hitch to solve.

The activity was based on role-playing and participant observation and was repeated in multiple sessions with different subjects. Entering into the details of the activity, the teamwork recreated a table reading—the sharing of a script—during which the participants had to identify themselves respectively in some actors and a director.

The director had the role of interpreting the script and assigning the roles of the three subjects (in both cases in which the experiment was repeated two boys and two girls were involved and the role of director was assigned once to the male designer and once to the female designer, so as to have fair results).

The researchers focused on the interpretative choices of the director, in order to observe the surfacing of unconscious bias by analyzing the dialogue between the participants, the choices that the director made and the consequent reactions of the other participants.

From some extracts, in particular, the tendency to be dragged by the inertia of thought arose, due to the lack of time or issues defined by the participants themselves as "marketing choices." Most of participants took refuge, more or less consciously, in stereotyped models of reality has emerged explicitly.

The experimental activity then flowed into a video that highlights these very aspects. Despite the fact that during the debate that followed, positions emerged that highlighted the open-mindedness of young designers, when we are put in the position of having to make quick design choices, it is *natural* to use those representative models that we have absorbed and that are now part of our visual and cultural baggage, which according to a mechanism of *new realism* we have assumed to be normal. In terms of learning, this type of activity allows students to become aware of the automatisms that we put in place when we are called upon to design. Through the immersive approach, students are allowed to experience the phenomena under study. The approach makes evident the mechanisms of which each of us believes to be free, increasing awareness of the stereotypical effects of common sense.

On the other hand, the visual outcome transposed into a video can be repurposed as a communicative artifact addressed to the community of designers themselves, in order to broaden the reflection in a perspective that places design students at the center of a system of responsibility. The students themselves are the protagonists of bottom-up actions that can broaden the debate and call peers and future designers to reflect, triggering a virtuous circle that opposes the vicious circle of stereotypes.

8.5 Conclusions

The activity carried out by the DCxCG research group within the university propose a proactive approach; the goal is to work upstream of the problem by raising new generations of designers capable of shaping an innovative media landscape from a socio-cultural point of view. A landscape characterized by inclusive and sustainable messages, in contrast to the androcentric worldview that pervades the *mediasphere* nowadays.

Referring to Dewey's point of view: the educator has the responsibility to foster cultural changes that are accepted, interiorized, and acted upon by students and which form small pieces collectively leading to social innovation. The fundamental components on which we insist can be summarized in two unavoidable points: (1) to equip students with tools for a critical reading of designing, enabling them to develop gender-sensitive thinking and apply it to their own design practice; (2) to encourage self-reflection, a practice that allows you to design consciously, governing the effects of meaning produced and avoiding running into the trap of stereotype.

The experiments carried out in the educational field can therefore be considered a response to the resolutions of the European Parliament which relate gender issues and the quality of media messages, placing the communication designer at the center of a complex system of responsibilities: "[the resolution] highlights the importance of promoting media literacy [...] so as to encourage young people to develop critical thinking skills and to help them identify and denounce sexist representations and discrimination" (2017/2210(INI)).

At the same time, these actions state the recognition of the issue by a public university institution, of national and international importance, and its assumption of a leading role within the design disciplines in order to recall the responsibilities that every educational path must assume to get to loosen the vicious circle of stereotypes, and to contribute to a cultural change that must invest society as a whole.

Notes

1 See: www.bbc.com/news/world-europe-56516462.
2 See: www.senato.it/service/PDF/PDFServer/DF/356433.pdf. *Ddl Zan* is a draft law against homotransphobia, which proposes measures to prevent and fight discrimination and violence on the grounds of sex, gender, sexual orientation, gender identity, and disability. The law was approved by the Camera in November 2020. The *ddl* was then scheduled for reconsideration on March 30, 2021, which the political party of Lega had rescinded. Senator Pillon assessed the *ddl* as "not a priority," despite the fact that recent months in Italy are marked by episodes of homophobia that have resulted in physical violence.
3 See: file:///Users/fra/Downloads/20190452_mh0419050enn_pdf.pdf. *Beijing + 25* is "the fifth review of the implementation of the Beijing Platform for Action in the EU Member States."
4 See: www.dcxcg.org/. The DCxCG group deals with research and educational projects to promote the social responsibility of Communication Design in the field of Gender Cultures. It aims to give its contribution in the development of critical knowledge and, at

a pragmatic level, in the design of actions to raise awareness, information, and training for the visual construction of the society, contributing to the development of communication projects for a redefinition of formats and rules of communication.

5 Zingale, S., "Immobili visioni" in Baule, G. & Bucchetti, V. (2012) *Anticorpi comunicativi*. FrancoAngeli, Milano.

6 It is named after a well-known Italian food brand that bases most of its communication on what is named the "traditional" family model, derived from patriarchal culture.

References

Anceschi, G. (1981). *Progettazione visiva: convenzioni e procedimenti di rappresentazione*. Edizioni Officina Immagine, Budrio.

Anceschi, G. (1992). *L'oggetto della raffigurazione*. Etas Libri, Milano.

Baule, G., Bucchetti, B. (edited by) (2012). *Anticorpi comunicativi*. FrancoAngeli, Milano.

Baule, G., Caratti, E. (edited by) (2017). *Design è traduzione*. FrancoAngeli, Milano.

Bucchetti, V. (2021). *Cattive immagini*. FrancoAngeli, Milano.

Bucchetti V., Casnati F. (2019). *The Contribution of Communication Design to Encourage Gender Equality*. In: (edited by) Ambrosio M., Vezzoli C., *Designing Sustainability for All*. Proceedings of the 3rd LeNS World Distributed Conference, April 3–5, 2019. Vol. 1, pp. 619–623. Milano. https://re.public.polimi.it/retrieve/handle/11311/1130363/492449/2019_LENSCONFERENCE.pdf

Bourdieu, P. (1998). *Il dominio maschile*, It. Trans. Feltrinelli, Milano.

Camussi, E., Monacelli, N. (edited by) (2010). *Questioni sul corpo in psicologia sociale*. Proceedings of Giornate di Studio GDG, May 7–8, 2010, Milano.

Capecchi, S. (2018). *La comunicazione di genere—Prospettive teoriche e buone pratiche*. Carrocci Editore, Roma.

Caratti, E. (2015). *Rimediazioni gender-sensitive. Contributi e progetti per la formazione di un immaginario consapevole*. FrancoAngeli, Milano.

Criado Perez, C. (2019). *Invisible Women: Exposing Data Bias in a World Designed for Men*. Vintage Publishing, London.

D'Amico, M. (2020). *Una parità ambigua. Costituzione e diritti delle donne*. Raffaello Cortina Editore, Milano.

Decataldo, A., Ruspini, E. (2014). *La ricerca di genere*. Carocci Editore, Roma.

Fabris, A. (2014). *Etica della comunicazione*. Carocci Editore, Rome.

Federici, E., Leonardi, V. (edited by) (2013). *Brinding the Gap Between Theory and Practice in Translation and Gender Studies*. Cambridge Scholars Publishing, Newcastle upon Tyne.

Maldonado, T. (1959). Educazione e filosofia dell'educazione. *Merkur*, Vol. 13, No. 9, pp. 67–77.

Mirzoeff, N. (2015). *How to See the World*. Penguin, London.

Mulvey, L. (1975). Visual Pleasure and Narrative Cinema. *Screen*, Vol. 16, No. 3, pp. 6–18.

Overall, N. C., Chang, V. T., Cross, E. J., Low, R. S. T., & Henderson, A. M. E. (2021). Sexist attitudes predict family-based aggression during a COVID-19 lockdown. *Journal of Family Psychology*. Advance online publication. https://doi.org/10.1037/fam0000834

Williams, M. K. (2017). John Dewey in the 21st Century. *Journal of Inquiry & Action in Education*, Vol. 9, No. 1, pp. 91–102. University of West Florida.

9

RESPONSIBLE TECH-FUTURES

Female artists as semantically sustainable tech developers

Boel Christensen-Scheel, Kristin Bergaust, Venke Aure, and Idun Alman-Kaas[1]

Reflecting on the role of gender in the sustainability discourse, the projects of several female artists stand out as material for discussion. Innovation discourse was, at least in the 20th century, driven by product orientation and market perspectives. Recent years innovation theories have increasingly been followed by sustainability perspectives, and circularity, re-use, focus on experience rather than product, as well as social dimensions of the production, have become familiar concepts. Still, the discourse is much focused on precisely the product or the function, and different aspects of sustainability relating to these products and functions. In contemporary art projects, we see instead that these functions and products are questioned, and other sides of the sustainability discourse are introduced. Aspects such as meaning, intention, ethics, context, and, one might also add, epistemology, are explored as part of a female take on technological and biological innovation systems and the ways these are displayed.

In this chapter, we will introduce four primary examples of female artists involved in bio-art and technology innovation that shifts the sustainability discourse and that might be said to promote a semantically responsible perspective, working with context and awareness, and where the meaning and ethical context of the project is as important as the products themselves: The art and innovation project *Human Fuel* by artist Hege Tapio, the sustainable art and technology projects of artist and researcher Rasa Smite, the "bio-art perspective" of professor and artist Laura Beloff and the artistic research project *Oslofjord Ecologies* by visual art professor Kristin Bergaust. In order to discuss these four examples, we will use theoretical perspectives from feminist technology development (Donna Haraway, Yvonne Volkart, C. Solfrank), and we will draw on ecosophical perspectives in philosophy and biology (Félix Guattari, Arne Næss, Maurice Merleau-Ponty, Gregory Bateson), environmental perspectives in art theory (Sacha Kagan, Venke Aure, Kristin Bergaust, Boel Christensen-Scheel),

DOI: 10.4324/9781003174257-11

FIGURE 9.1 *Pond batteries* by Rasa Smite and Raitis Smits, 2018, courtesy of the artists.

and anthropocene philosophy (Rosi Braidotti, Timothy Morton). Following Rosi Braidotti's and Sacha Kagan's art environmental theories, the gender perspective in the sustainability debate is also a culture producing mode of reflection transcending gender and tightly connected to the way we envisage the societal values and potential for development.

9.1 Is there a gendered tech development?

The history of the exclusion of women from the scientific and the art field is well known, and it is obvious to refer to the question Linda Nochlin raised in her essay from 1971 that goes "Why have there been no great women artists?" Through this question, Nochlin showed the structural exclusion practiced by the art museums exhibition practice. Way presents the exhibition *Art and Technology* that Los Angeles County Museum of Art (LACMA) arranged during 1971 as a clear example of this neglect of women and technology in the art field at this time (Way, 2016). The exhibition that presented the work of 16 artists did not include any woman at all. This resulted in the formation of the Los Angeles Council of Women Artists (LACWA) and a storm of protests, as here from *The L.A. Free Press*:

> We were particularly provoked by the blatant discrimination against women in the highly publicized Art and Technology show new at the Los Angeles County Museum of Art. Sixteen artists are represented in this invitational show – NONE are women, NONE of the technical advisors with whom they collaborated were women.
>
> *Way, 2016: 186*

From this time, feminist historians have focused upon how gender and technology impact one another through time. As a concretization of this, Way uses Judy Wajcman who claim that the gendering of technologies affect "the entire life trajectory of an artifact" in addition to the design and the multiple points of use of it (Way, 2016: 183). This consciousness regarding a male "determinism" that kept woman out of the field of art and technology and an understanding of technology characterized by a relative neglect of software led to a set of techno-feminist positions, a term Wajcman already introduced in 2004. These positions represented a new philosophy that implied wider contexts than the firmer representations of technology and art. From now on, we see the will that especially Donna J. Haraway has insisted upon in her texts: to build down dichotomies between biology and culture, the human and the machine, technology, and context (Haraway, 1991). A feminist epistemology that emphasizes contexts, networking, processes, activism, and critical investigations became an alternative techno-feminism. This alternative and extended technological epistemology also include what Haraway calls "art science activist worldings" and projects connected to ecology has to a great extent become a part of this feminist philosophy (Haraway, 2016: 76).

Now asking the question whether there is a gendered technology development, we bear in mind these perspectives from the art and technology history: women as technology developers are rising, but like artists, they have in both near and far history not been acknowledged as such to the same extent as men. Leaning on recent innovation theory, it becomes an asset to include otherness, that is, those voices that are not already represented in the decision-making, in order to *reach wider audiences* and represent a greater public. It also becomes a goal to be *sustainable* or *ethical* in the innovation choices made, in particular in regard to production. Perhaps can female tech developers with a creative or artist connection contribute to other and more humanistically founded tech changes? The combination of the human body and machine, or the living organism and the technology, is at least one such area of immense contemporary importance, that might carry a gendered perspective (Braidotti, 2006). As the female bodily experience is other than the male, there might be other perspectives to explore and other awarenesses to encounter. Gender research in medicine has, for example, shown that through the fact that the main number of cardio researchers were men, the knowledge on the female heart and heart conditions remained low for a long time. In a similar way, living technology research and human-technology innovations requires a female or gendered perspective – even if we strive for "principal equality" (Næss, 1989), we have to acknowledge the actual and embodied gendered experiences and differences (Braidotti, 2006). In the end, this is a fundamental ethical perspective in technology development and innovation cultures, where both modes of production and manifestations of systems, are exerting power and practice precedence that never should be taken for granted. This is the epistemology of the tech cultures and "innovation paradigm" that we here seek to unravel, with the help of four artists/innovators. All practice contains theory, and the relation between "what" we need and "why" we need it represents the core of sustainable culture practice and development.

FIGURE 9.2 Hege Tapio in bio-tech conference in Norwegian University of Life Sciences, courtesy of the artist.

9.2 Human fuel

The project *HumanFuel* by Hege Tapio:

> HUMANFUEL is presented as a response to the hunt for alternatives to fossil fuel and posits an obligation to embrace a larger perspective to the Anthropocene – to include the human form in the biological chain of recycling – through the proposal of using human fat to fuel vehicles.
>
> *Tapio, 2016: 1*

At the Hybrid Matters exhibition in 2016, she proposes to use the extracted human fat from liposuction to produce biofuel for cars and transportation. In order to explain her concept, she herself underwent a liposuction prior to the exhibition and presents herself as both the scientist/artist in the project and the private source of energy. The project connects two central issues in contemporary times – the context of the vanity industry and the current climate issues and need for new, sustainable energy sources.

The proposition to use the leftover fat from the cosmetic industry makes sense, as it takes advantage of a bi-product or waste product from something else. At the same time, the critical, artistic perspective is instantly conveyed, as the use of "waste" from the human body is considered immediately problematic. Also, the ethical problematic sides are evoked, as the human body usually is lifted above industrial use. The value of the body carries a certain symbolic value, but as Tapio illustrates with her project, what this symbolic value is in a cosmetic surgery industry is highly arguable and the use of this fat for a sustainable purpose, thus seems the most ethical proposition. Whether this would be economically viable is a financial question, but the idea in itself can be considered as both innovative and culture critical at the same time. The value of the female body is directly connected to sustainability issues, showing both vulnerability and responsibility from a female perspective. The female body is already industrialized and objectified, but as users and buyers in a cosmetic industry, women are also subjects and game-changers.

The Latvian laboratory Lipo-technica simulates an Eastern European tech company that is willing to take on this human fuel challenge. Tapio as artist, scientist, and woman is this context is emerging as both technology developer and reflexive practitioner, opening up for ethical discussions and considerations much needed in today's tech industries.

9.3 Pond batteries

Rasa Smite with her partner Raitis Smits also combine academic and artistic backgrounds, evolved in close proximity to their activities connected to the media art center RIXC of which they are the founders. They both work as organizers and curators in the field of art, technology, ecology, and educators on a high level of artistic research.

Their ongoing artistic research with electricity-producing bacteria has had a range of manifestations. A common denominator and driving idea for these projects is the ambition and interest in having the natural forces of the bacteria become visible or audible in unexpected ways. How can you manifest the invisible processes of energy in nature? One example is the time-lapse video of an installation of bacteria batteries in a pond in a botanical garden. Over some months, the installation produces a display of electricity in numbers while we follow the climatic and seasonal changes and see snow falling and ducks disturbing the batteries. The measurements of science are transformed into a poetic observation, rather than a utilitarian demonstration.

9.4 Bio-art innovation

The "bio-art perspective" of professor and artist Laura Beloff concerns an increasing interest in scientific research concerning the "physiological and psychological impact" of natural environment to human health and well-being, suggesting that green environment is beneficial and has restorative qualities for human health (e.g. research by R. Mitchell, F. Popham, T. Hartig, S. Kaplan, R. Ulrich, L. Tyrväinen

FIGURE 9.3 (a and b) [side-by-side] *Human Fuel* by Hege Tapio, Hybrid Matters exhibition, Kunsthall Grenland 2016, courtesy of the artist.

FIGURE 9.4 Another version of *Pond batteries* by Rasa Smite and Raitis Smits, 2018, courtesy of the artists.

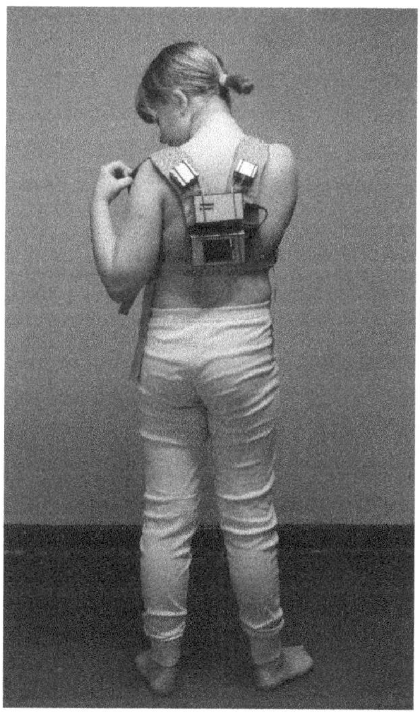

FIGURE 9.5 Wearable technology by Laura Beloff, courtesy of the artist.

among many others). In her 2013 thesis *The Hybronaut and The Umwelt: Wearable Technology as Artistic Strategy*, she:

> explores the use of irony in networked wearable technology art as a strategy to emphasise the complexity of conjunction between techno-organic human and the techno-organic world. The research addresses the relationship between technologically enhanced human and networked hybrid environment, and speculates on the impact of technological enhancements to the subjective construction of Umwelt through ironic interventions.
>
> *Beloff, 2013: 3*

Laura Beloff is an internationally acknowledged artist who is known for working in the interdisciplinary field between art and science. Beloff's role in the scientific field is as a researcher who functions as an academic leader with the core in arts and uses methods such as artistic research. In the projects, she typically collaborates with other scientist that are specialists on their fields of knowledge. Her work has been exhibited in museums and galleries across the world, and most recently in Denmark, Norway, Belgium, Finland, Germany, Austria, Brazil, and Russia (Beloff, 2020).

Beloff works within the cross-section of ecological-scientific-technological-biological scene of artistic and practice led research. In this practice, she also raises awareness to the ethics and philosophical questions concerning technologically manipulation of living organic matter. In *The Fly Printer* (2014–2016), Beloff collaborated with María Antonia González Valerio. Together they developed an organic printing system of fruit flies. The first version of *The Fly Printer* (2014) gave the fruit flies different food immersed with laser jet ink colour. The flies were contained in a glass bowl and underneath the glass was white paper. When the flies digested the ink-food, they became printers, leaving randomized coloured dots on the paper. In the latest version of *The Fly Printer*, Beloff and González added an A.I. level to the project. As the flies were printing, a machine vision system was observing the imagery; then the A.I. convolutional neuronal network (CNN) analyzed and found patterns from the images and interpreted it to a human observer.

The Condition – cloned Christmas trees (2016–) is an ongoing project with Beloff in collaboration with Jonas Jørgensen. The project is a research into the resilience and perseverance of a Christmas tree from the Danish tree *Normann Fir*, which originates from seeds that has been collected from the Caucasus forest. The Danish version is cloned to become the perfect christmas tree, so this is no longer a natural tree. The trees in this installation is the same genetically tree. They are aligned in a matrix structure, hung on the wall 4×3 in a rotating box. There are several growth lights mounted on the installation directly pointed at the small trees. *The Condition* is described as an exploration of an open-ended physical alteration process, deriving human intervention and technological evolution in conjunction with natural forces (Hybrid Matters, 2016).

9.5 Three or more ecologies

The artistic research project *Oslofjord Ecologies* lead by professor and artist Kristin Bergaust. An investigation of environmental strategies through artistic research methods in the fjord of Oslo. The project gathers diverse artists and researchers in order to generate different modes of knowledge production in relation to sustainability issues.

Kristin Bergaust, artist and professor in contemporary art at the Oslo Metropolitan University, has throughout her practice been active in art and technology networks. Since 2008, Renewable Futures Network with Rasa Smite and others have been important collaborators and peers. Oslofjord Ecologies, an artistic research project initiated by Kristin in 2016, was connected to Renewable Futures through Creative Europe financing and a Nordplus Horizon collaboration. The initial transdisciplinary workshop in November 2016 involved contributions from around fifty participants, artists, scientists, other professionals, and students. Since then ideas, questions, and productions have evolved through exhibitions, performance, and activity programs (Oslo 2017 and Helsinki 2018). Some of the projects involved have grown into mature doctoral-level research, and art works have become more consistent and permanent presences. Most importantly, the complex thinking and discussion is documented, recorded, and re-discussed in order to produce new reflection layers on top of the reflected materiality that is produced – an anthology of the project was published in the 2020 anthology, *Oslofjord Ecologies – Artistic Research on Environmental and Social Sustainability*.

Oslofjord Ecologies was in its conception an amalgam of artistic and academic interests encompassing research of many natures: artistic, experiential, scientific, pedagogical, historical, and cultural. Artistic perspectives and sensibilities were broader than the techno-ecological field. The shared perspective was ecology in an open sense, resonating with Felix Guattari's essay *The Three Ecologies* from 1989 (Guattari: 58) where he calls for new articulations of "a nascent subjectivity, a constantly mutating socius, an environment in the process of being reinvented," often summed up as layers concerning the mental, the social and the environmental. In order to formulate a research field of art and ecology that could be artist-driven, but embrace knowledge from multiple competences and sources, the theoretical definitions were left open beyond these quite demanding perspectives. Through this approach, participating artists questioned art in the context of ecology: What does art do? How can artists contribute? What kind of knowledge does artistic research produce? Rather than being answered, these questions continued motivating ensuing reflections and inspiring new work modes. During the next developments, ideas such as Donna Haraway's concepts of "becoming with" and sympoiesis, the possibility of cocreating our lives and circumstances with and in consideration of other beings have become important for concepts of interspecies relations. Discourses of new materialism and queering strategies that obstruct the dichotomies of human and nature are also inspiring a range of practices from pedagogy to artistic research practice and design.

9.6 An eco-feminist bio-tech agenda?

The four examples we have presented clearly illustrate that technology development and art are situated terms, framed by social, cultural, and gender related perspectives. The examples show women strongly intervening with technological developments in art, this area that not long ago was noticeably male dominated. Regarding to this fact Jennifer Way refers to how Grant Taylor in his book *Up for Grabs: Agency, Praxis and the Politics of Early Digital Art* from 2013 claims that women have been largely ignored in the formation of early digital art, an aspect he relates to a construction of history-making through a masculine discourse (Way, 2016: 186). Women have been present in recent avant-garde art development, but we still see that numbers of solo exhibitions and value in money are dominated by male artists. All four examples show women driven by engineerial curiosity, however, undertaking their developmental activities with a concern for what development means and engenders. In particular, the artistic perspective contributes to a reframing and contextualization of both tech development and sustainability. For example, the connection of the human body fat directly to the fuel problem in relation to sustainability that Tapio sets up, creates a critical awareness about our priorities, showing both a completely sustainable model, but also directly absurd view on purposefulness. To use human body fat and waste from the cosmetic industry is a definite possibility; at the same time, we ask ourselves the question about how it would be to drive around town with our own or others' body fat as fuel. The idea is innovative and repulsive at the same time.

Laura Beloff conveys some of the same critical awareness in her projects, where tech development is accompanied by ways of showing and seeing this development. One could say that the experience of "being seen" as a woman, accompanies the feminine tech production. How is it seen? How can it be seen? To view things in different ways represents a scientific theoretical and epistemological take onto the tech activity, an epistemological awareness promoted by Venke Aure in her concept of *art didactics* (Aure, 2006, 2011). Such an epistemological awareness can be brought on by these gendered or artistic perspectives. Yvonne Volkart suggests the term *techno-eco-feminism* to convey this interplay between ecological and techno-feminist aspects (Volkart, 2020). Feminism has many positions, but in the sense of Volkart and Haraway, techno-feminism underlines the "equal" opportunities that lie in technology. Tied to the body, but also superseding the body, technology can overcome gendered perspectives and become a strengthening of the body. Technology may provide you with an extra set of arms or legs; it may strengthen your eyesight or your hearing and create a different appearance or make appearances indifferent or invisible. Technology in this sense, and in the sense of philosopher Jacques Rancière, create new landscapes of the sensuous, that supersedes, alters, or fuses bodies. This is a techno-positive attitude, but it is not to say that we do not see many of the same social problems in the sphere. Bourdieu's concept of social reproduction is also present in the technological era. However, the techno-feminism promoted by Haraway and Volkart suggest that technology contains further opportunities for inclusion and gender balancing that we have not yet explored. Tech development then needs to be accompanied by repeated stages of reflection.

Today, these reflexive modes imply sustainability concerns, as the technology development can never be detached from its context of *use* (and *mis-use*). In *The Craftsman*, culture sociologist Richard Sennett describes Oppenheimer's invention of nuclear technology (Sennett, 2009: 1). He very vividly conveys the intentions of the scientist, technologist tied to development and concern with a specific problem that needs to be solved. Also, Sennett shows how the ideas of progress and development, construction, and technological achievements can never be detached from their societal contexts and use. Head and hand must work together, he says. Following Volkart and Haraway then, the technology in some way substitutes the cultural and performative pre-conditioning of the gendered body and create hands and/or minds freed from the subject history. If this points toward a brighter or a more obscured future in terms of ethics, is yet to see.

9.7 Art as part of tech development futures

Latvian artist Rasa Smite's projects explore a more media oriented and structural approach. As a social science PhD and EU researcher, she creates networks but remains close to natural environment and environmental research in her RiXC New Media Center activities. Her "Pond Batteries" and "Pond Radio" aim to show natural activity and connections between biological processes and technological processes. The movements and processes of the pond bacteria that result in energy production are a natural process; however, it constitutes a possible valuable technology that also acknowledges the environment. The art and media sides of her projects constitute a communicational side; we get to hear or see processes and energy release that cannot be spotted by the mere eye. These visualizations also make us able to reflect on what is going on – we understand the technological process and can take a stand in relation to it. An important side of the artistic contribution to tech development is thus not only ethics and reflections, but the ability to communicate and visualize processes for us to understand what is going on and why (Christensen-Scheel, 2020).

Entering more recent modes of production, it seems that these levels of reflection and awareness of context, use, and ethical considerations still are aspects that need to be inextricably tied to design and engineering. There are billions of possible solutions to our problems; not all of these solutions entail production, but all of them needs reflection. Techno-eco-feminism in this context represents reflectional strength – it is naturally not only about women; it is about ways of seeing and considering possible futures, physically, mentally, and socially. Here, we can connect to Fèlix Guattari`s "three ecologies" – the environmental, the social, and the mental ecology. Guattari was a clinical psychiatrist and philosopher, who observed the relation between "inner" and "outer" human states. In a 1970s and 1980s, growing focus on pollution and preservation of nature, he attempted to make the connections between the human treatment of nature and the human nature of mind.

Through Guattari then, the logic of mind and the logic of nature is cobbled – the way we treat ourselves, the way we treat others, and the way we treat our environment are not separate things, but intertwined parts of existence. If we abstract the idea sufficiently – we can think of our minds as nature as well, where the chemical substances of the brain and the primitive modes of behavior are shaped by biological factors and interconnected with larger ecosystems. Physicist Gregory Bateson was making similar connections – between the biological systems and the structure of the mind (Bateson, 2000). As complex as the urban human behaviors are, our feelings and choices are guided by both predictable and unpredictable patterns as in ecosystems. The choices of nature and the choices of humans might be more alike than we are aware of. Gender in this concern is a biological construction, but it is also a power issue, an aspect of dominance and of territory. Introducing modalities of ethics based on precisely interconnectedness and balancing of power seems to be a necessity then, in order to reduce "natural inequalities" and different unfortunate consequences of social reproduction (Bourdieu, 1993; Braidotti, 2006). Sustainability when it is not only about nature and environmental issues is social and mental as well (Guattari, 1989). Therefore, we now also talk about *social sustainability* (Bergaust, Smite & Silina, 2020), where our social systems are closely linked with our ability to act eco-ware.

Through our examples and more recent theory: Ecosophical perspectives are here not only tied to a *preservational* mode but are an ethical-aesthetical attitude applied to development and production. With ethical-aesthetical is meant a combination of physical surroundings, modes of production with an ethical and epistemological consideration (Christensen-Scheel, 2012). Technological development is then not primarily a productive mode; it is also reflective, as underlined by Sennett (2009). Artists have always developed new techniques, materials, and methods of production but have always also dealt with the semantics of the production. Semantics is the production of meaning that accompanies the product or production – and the layers of meaning introduces a reflexive mode as well as deals with aspects of ethicality and criticality (Rogoff, 2006). Art, and female artists, may then very well produce necessary dimensions of future tech development, tied to an ethic-aesthetic awareness.

9.8 A gendered sustainability?

In a second phase of her research, Braidotti focuses on the posthuman condition in three publications. Her trilogy consists of *The Posthuman* (2013), *Posthuman Knowledge* (2019), and *Posthuman Feminism* (to be published, held as lecture), which concentrates on the discourse concerning advanced technologies on one hand and the advanced environmental destruction on the other. The question on what it means to be human is getting more unclear, as the boundaries between human and the other becomes more and more blurred out, through our digital lives, reproduction technologies, and geno-modified foods. The development in the technological world is taking us further from our original human and organic sphere. This may result in the need of establishing a new definition on what is to be human. Braidotti looks at this development and searches for new ways for us to produce knowledge about these possible thoughts of forthcoming human-techno societies which also implicates the comprehensive theme ecology.

So what is our perception of technology today? Technology can be thought of as a means to any purpose – a skilled method in some way, of achieving a certain action or product. Living in a tech-centered research world, we question the sometimes-naive belief in technology as a solution, the tech romantic visions. However, technology can also be seen as the mere prolongation of our bodies, and it is always tied to our capacity to imagine other and better realities for existence. Or a way to dissolve natural differences and prejudices that may cause inequalities. So, we might ask, is our ability to imagine through technology dependent on our bodies? On our existential experiences? It is beyond questioning that medical technologies have been developed slower for the female body and that female diseases have been underprioritized in research. Will the female future be threatened then, because the majority of tech developers, tech researchers, and tech managers are men?

All of these are rhetorical questions, some more factual than others – however, the point made by Braidotti (2006), and that we know from the sociology of Bourdieu (1993) and numerous statistics, is that if we do nothing, the existing power structure will most likely reproduce itself. It will do more of the same. We can then move on to ask what gender is today? Outdated categories? Perhaps, but ethics and representativity which closely follows the gender and feminist discussions are not outdated in the way suggested above. Feminist discussions and queer theory have used a lot of space discussing a wide range of body identities, who may be defined in the gender categories, or that might exceed them or blow them up. Gender in a performance perspective (Butler, 1993) is performed, and produced (Beauvoir, 2005). Gender is projected onto individuals as part of different cultural paradigms, and performed as part of subcultures, career tracks, and socio-cultural strands. But gender is also biology and nature, something that exceeds our mere will and influence.

When we look at sustainability issues, we might ask then, who identifies with them. Female artists or male technologists? Male artists or female technologists? We have used prominent female theoreticians to explain our perspective here as well as radical and experimental male philosophers, because they are the ones who have felt the precariousness to address the issue, to present ethicality and representativity in technology development – dwelling on the possible embodied challenges the future digital paradigm will contain.

Having presented a selection of female tech-oriented artists driven by sustainability issues, we present a gendered perspective onto the tech innovation debate. Interestingly, the relation between art and technology has always been close, in the way that art in some way is technology, at least usually presented in or through one. When we think of the future, we want to imagine responsible developers or changemakers, that are both able to take us further, enhance our life modes, but at the same time being aware of who falls outside of the frame. The artists perspective can let itself be driven by explorative concerns but can also address power issues and sustainability. Art as an autonomous field of production is a way of questioning our routines, by not having to prove profitable or achieving particular goals (Christensen-Scheel, 2013). The four female artists presented here each give their take on technological development, but as we see it, they represent a technology of "semantics" that supercedes the mere productional mode and makes a bridge over to the field of semantics, where technology is also always tied to meaning and whilst developing our future, these critical and reflexive attitudes that give another picture of the story being told, may very well be the ones that make sense.

Interestingly, the relation between art and technology has always been close, in the way that art in some way is technology, at least usually presented in or through one. When we think of the future, we want to imagine responsible developers or changemakers, that are both able to take us further, enhance our life modes, but that at the same time are aware of who and what falls outside our meaning hierarchies. The artist perspective can let itself be driven by explorative concerns but can also address power issues and contextualize sustainability. Art as an autonomous field of production is a way of questioning our routines, by not having to prove profitable or achieve particular goals (Christensen-Scheel, 2013). The four female artists presented here – Hege Tapio, Laura Beloff, Raisa Smite, and Kristin Bergaust – each produce a technological development that at the same time comments on what sustainability is. As we see it, they represent a technology of "semantics" that supersedes the mere productional mode and makes a bridge over to the field of communication and value production. Here, technology is also always tied to meaning, and whilst developing our future, these critical and reflexive attitudes give another picture of the story being told, and in the future, they may very well be the ones that make sense.

Note

1 This article is part of the Future of Living Technologies project led by Kristin Bergaust and financed by DIKU – Direktoratet for internasjonalisering og kvalitetsutvikling i høyere utdanning 2020–2024 FeLT (Futures of Living Technologies) OsloMet.

References

Aure, V. (2006). *"Formidling av bilder til barn – som kunstdidaktisk diskurs." Tidsskrift for børne- og ungdomskultur Nr 51 / Kunstformidling for og med børn og unge.* Odense: Syddansk Universitetsforlag & Børne- og Ungdomskultursammenslutningen.

Aure, V. (2011). *Kampen om blikket. Academic Dissertation for the degree PhD.* Stockholm: Stockholm University.

Bateson, G. (2000). *Steps to an Ecology of Mind. Collected Essays in Anthropology, Psychiatry, Evolution, and Epistemology.* Chicago: University of Chicago Press.

Beloff, L. (2013). *The Hybronaut and the Umwelt: Wearable Technology as Artistic Strategy.* PhD thesis, University of Plymouth Research Theses, PEARL Home (plymouth.ac.uk).

Beloff, L. (2020). Research and artistic practice. Available at: realitydisfunction.org
Beloff, L. and Jørgensen, J (2016), 'The Condition', Hybrid Matters, Kunsthall Grenland, Available at: hybridmatters.net
Braidotti, R. (2006). *Transpositions*. London: Polity Press.
Braidotti, R. (2013). *The Posthuman*. London: Polity Press.
Braidotti, R. (2019). *Posthuman Knowledge*. London: Polity Press.
Beauvoir, S. de (2005). *Det annet kjønn*. Oslo: De Norske Bokklubbene.
Bergaust, K., Smite, R., and Silina, D., eds. (2020). *Oslofjord Ecologies – Artistic Research on Environmental and Social Sustainability*. AS.Vol.18-OSLOFJORD-ECOLOGIES-download.pdf (rixc.org).
Butler, J. (1993). *Bodies That Matter*. London: Routledge.
Bourdieu, P. (1993). *The Field of Cultural Production*. Cambridge: Polity Press.
Christensen-Scheel, B. (2012). The Ethic-Aesthetic Way of Wonders. *Nordic Journal of Art and Research*, *1*(1), pp. 19–38. https://doi.org/10.7577/information.v1i1.215
Christensen-Scheel, B. (2013). Application and Autonomy: The Reach and Span of Contemporary Art Didactics. *Nordic Journal of Art and Research*, *2*(2), pp. 109–123. https://doi.org/10.7577/if.v2i2.729
Christensen-Scheel, B. (2015). Recognizing the Observer of Art. In: Aure, Venke; Bergaust, Kristin, eds. *Estetikk og samfunn. Tekster mellom samtidskunst og kunstdidaktikk*. Del 1, artikkel 1. s. 37–56. Bergen: Fagbokforlaget.
Christensen-Scheel, B. (2020). ECO-Art and Awareness: Sustainability Through Artistic Research. In: Bergaust, K., Smite, R. and Silina, D., eds. *Oslofjord Ecologies – Artistic Research on Environmental and Social Sustainability*. AS.Vol.18-OSLOFJORD-ECOLOGIES-download.pdf (rixc.org), pp.25–36.
Guattari, F. (1989). *Les trois écologies*. Paris: Galilée.
Haraway, D. (2004). *The Haraway Reader*. London: Routledge.
Haraway, D. (2016). *Staying with the Trouble. Making Kin in the Chthulucene*. Durham and London: Duke University Press.
Haraway, D. (1991). "A Cyborg Manifesto: Science, Technology, and Socialist- Feminism in the Late Twentieth Century," in *Simians, Cyborgs, and Women: The Reinvention of Nature*. New York: Routledge, pp. 149–181.
Kagan, S. (2011). *Art and Sustainability: Connecting Patterns for a Culture of Complexity*. Germany: Transcript-Verlag.
Keitsch, M. and Vermeulen, W.J., eds. (2020). *Transdisciplinarity for Sustainability*. London: Routledge.
Merleau-Ponty, M. (2003). *Phenomenology of Perception*. London: Routledge.
Morton, T. (2018). *Being Ecological*. London: Penguin Books.
Nochlin, L. (1971). "Why Have There Been No Great Women Artists?" *ARTnews* 9, pp. 22–39.
Næss, A. and Rothenberg, D. (1989). *Ecology, Community and Lifestyle*. Cambridge: Cambridge University Press.
Rogoff, I. (2006). "Academy as Potentiality." in A. Nollert et. al. (red.), A.C.A.D.E.M.Y. Frankfurt am Main: Revolver.
Sennett, R. (2009). *The Craftsman*. London: Penguin Books.
Soini, K. and Dessein, J. (2016). Culture-Sustainability Relation: Towards a Conceptual Framework. *Sustainability*, *8*, 167; doi:10.3390/su8020167 www.mdpi.com/journal/sustainability
Sollfrank, C., ed. (2020). *The Beautiful Warriors. Technofeminist Praxis in the Twenty-First Century*. Brooklyn, NY: Autonomedia.
Tapio, H. (2016). "Human Fuel," artists statement https://tapio.no/wp/humanfuel/

Taylor, G.D. (2013). "'Up for Grabs': Agency, Praxis, and the Politics of Early Digital Art." *Lateral* 2. http://lateral.culturalstudiesassociation.org/issue2/theory/ taylor (accessed October 2013).

Volkart, Y. (2020). Techno-Ecofeminism. Nonhuman Sensations in Technoplanetary Layers. In: Sollfrank, Cornelia, ed. *The Beautiful Warriors. Technofeminist Praxis in the Twenty-First Century*. Brooklyn, NY: Autonomedia (pp 111–135)

Way, J. (2016). Digital Art at the Interface of Technology and Feminism: Paul/A Companion to Digital Art. Mars 2016, www.researchgate.net/publication/313720667

PART III
Infrastructure

10

CHANGING TRADITIONS

The case of Amazonian riverine women and Belo Monte hydropower dam

Satya Maia Patchineelam and Maartje van Eerd

Country: The case study is in the state of Para, in the northern part of Brazil.

10.1 Introduction

Riverine communities from the Xingu River that is found in the Brazilian Amazon had to leave their homes alongside the river due to the construction of the Belo Monte hydropower dam. They were resettled to Collective Urban Resettlement sites (CURs) which were found on the outskirts of the city. The riverine communities were distributed to five different CURs, which have affected people's social connections, separated families and family members, distanced them from their livelihoods, primary income resources, and have jeopardized their traditions.

Riverine people, also known as "*Povos das Aguas*" – People of Waters, are dependent on natural resources that are found in the forests and rivers, and on subsistence economy relying on their extended families and community. From the project plan to the construction of the dam, the riverine people were not treated any differently to other members of the population, and their traditions were not taken into consideration, which resulted in placing them in the rural and city dwellers' category (Energia, 2011). Researchers have affirmed that the riverine folk are the most affected people when there is a development project; this is due to their isolation and lack of access to public policies (da Costa Silva et al., 2010).

Most of the riverine families from the Xingu region are known to have two houses, one on the riverbank and another alongside the streams in the city center of Altamira. During the resettlement process, most of the families lost both houses due to the formation of a reservoir that flooded the houses on the riverbank and the houses on the stream found in Altamira. The people were resettled to resettlement sites that have detached them from the river, and this has unfortunately marginalized the women, isolated them from society and disconnected them from their traditions.

DOI: 10.4324/9781003174257-13

The new living locations have also separated them from their source of livelihood which has consequently affected the riverine traditions and reinforced their subordinate position within the household, which has resulted into inequality between genders by limiting women's opportunities and pressuring them to become more dependent on men.

This chapter focuses on riverine women in the Brazilian Amazon who, due to the dam construction, have been forced to move away from the river to be resettled to a Collective Urban Resettlement site, away from their previous land, disconnected from the river, and from their livelihoods. This involuntary relocation from their original location has impacted their lives in many facets which include social connections, their roles in the household, and their traditions.

The focus will be on how riverine women's lives changed after resettling to the outskirts of the city of Altamira which answered the research question: *How has the construction of Belo Monte dam in the Xingu River changed riverine women's livelihood traditions?* The first section includes a short overview of the academic debate on resettlement from a global perspective, and then zooming into resettlement related to dam construction. After that the Brazilian context is reviewed. The second section provides a brief explanation of the Amazonian riverine population and their traditions. The third section looks specifically at the context of Belo Monte and the resettlement of the local riverine community. The fourth section outlines the methods chosen for this research, and the last section before the conclusion focuses on how this shift has affected riverine women's lives and traditions, and finally the last section is the conclusion.

10.2 Development-induced displacement and resettlement

Resettlement disrupts family life and the social fabric of those displaced, and often negatively impacts those affected. The main causes for resettlement are development, conflict, and climate change. The World Bank describes resettlement as social fabricy which those adversely affected [by development projects] are assisted in their efforts to improve, or at least restore, their incomes and living standards' (World Bank, 2015).

In recent decades, Development-Induced Displacement and Resettlement (DIDR) has gained its scale and impact due to rapid urbanization, population growth, and a boom in infrastructure development (Terminski, 2015; van Eerd, 2017). In 2018 Cernea and Maldonado reported a conservative estimate of 20 million per year or 200 million displaced between 2011–2020 to be affected. This not only causes stress and trauma physically, economically, and socio-culturally but also gives rise to what Cernea (2000, p. 3660) famously referred to as "impoverishment risks."

Resettlement involves displacement, which is described by Cernea (2006) as "people losing their land, house, or both." It is important to recognize that because it has an enormous impact on those affected. Regardless of what happens after that, the impact of being displaced from the place where they were living, where they were connected to the land and the surroundings, in terms of work, education, access to services, having a social network to rely on and having a history in that particular place, is crucial (Papmehl-Dufay, 2015). Resettlement combines displacement and [should include] livelihoods restoration (Asthana, 1996; Terminski, 2013). It constitutes spatial and social elements as it needs to include appropriate planning and support systems in the new location (Muggah, 2008; Terminski, 2013).

Displacement and resettlement can be both voluntary and involuntary; those who are involuntary resettled tend to feel even more anxious and insecure (Asthana, 1996), although it is not very easy to distinguish between voluntary and involuntary resettlement as sometimes what is presented as voluntary resettlement is experienced by communities as being impossible to refuse and thus involuntary (van Eerd, 2008).

Resettlement is known for its overall negative impact on affected communities who generally tend to be worse off after they were forced to move (Cernea, 2000, 2009; de Wet, 2006; Mathur, 2016; van Eerd, 2018). The affected people often grow deeper into economic and social marginalization which is often due to poor project planning (Cernea, 1997). Marginalization is a characteristic of resettlement when there has been a loss of income, social disruption, and the fragmentation of community (Cernea, 1997).

10.2.1 Dams, displacement, and resettlement

By the end of the 20th century, there were more than 45,000 hydroelectric dams, which have displaced about 40–80 million people worldwide (World Bank, 2004). According to the International Displacement Monitoring Center (IDMC), an estimated 80 million people have been displaced by dam projects worldwide, and although their fate is largely unknown, evidence has shown that those affected tend to become impoverished and marginalized (2017).[1] In India, 60 million people were displaced due to dam projects between 1947 and 2000 (Samling, Ghosh et al., 2015). In China, the construction of the Three Gorges Dam (1993–2009) alone displaced 1.3 billion people (Zaman, Nair, and Guoqing, 2022, p. 3) Institute Igarape (2018) pointed out that from 2000 to 2017, more than 7.7 million people were displaced in Brazil, 17% due to development projects. According to Movimento dos Atingidos por Barragens (2004), 70% of the people who were displaced by dams were not recognized as impacted. Resettlement outcomes in Brazil have tended to mirror global trends, with those affected suffering impoverishment, marginalization, and other adverse impacts.

Impoverishment remains a legacy of displacement from large dams, which suggests that there are fundamental flaws in the way involuntary resettlement has been carried out (McDonald-Wilmsen and Webber, 2010).

10.2.2 Gender impact and resettlement

Impoverishment is experienced with different intensities by different people (Cernea, 2000; Koenig, 2009), as also shown by the fact that in most cases, women experience a more severe impact of resettlement (Terminski, 2015; Sikka and Mathur, 2018, Bajpai and Gautam, 2018, p. 93). Therefore, resettlement has become a factor in weakening the already heavily disadvantaged position of women (Terminski, 2013).

Women tend to experience shrinking livelihood opportunities, since having been uprooted they face greater difficulties in their new environments and need to build a new everyday routine from scratch (Smyth, Steyn, Esteves et al., 2015; Sikka and Mathur, 2018).According to Bajpai and Gautam (2018), most women lose their steady income as a result of DIDR. The displacement of women from managing and controlling the land, water, and forests destroys women's productive activities (Shiva, 2016). Losing access to common property resources and access to the forest threatens women's income (Sahoo and Mishra, 2016), thereby increasing women's dependence on men (Mehta, 2009) and increasing the inequality gap between these two genders.

From the social perspective, resettlement undermines social networks and causes disruption of family ties, resulting in the shrinking of the space a woman enjoys (Bajpai and Gautam, 2018). Maldevelopment, as explained by Shiva (2016, p. 5), is the "violation of the integrity of organic, interconnected and interdependent systems, that sets in motion a process of exploitation, inequality, injustice and violence," disharmonizes the relationship of men and nature, and men and women. This results in the fact that men are placed above women and nature, separating both and strengthening the patriarchal culture (Shiva, 2016).

10.2.3 Belo Monte dam, riverine people, and their resettlement

The Belo Monte Hydropower dam is in the Xingu River, in the Amazon Forest. It is currently considered one of the largest dams in the world (Chaves et al., 2019). A dam that has three decades of history, from negotiation of the organization of the biggest indigenous meeting in the history of the country in 1989 against the dam, which forced the project to be altered, to finally becoming the main development project under the former Brazilian president Luis Inacio Lula da Silva's agenda (Hall and Branford, 2012). It had a reservoir area of 516 sq. kilometers , displacing 1,241 properties from its former riverbanks and islands and 4,760 properties in the urban area (Energia, 2011).

The riverine, "people of waters," community in the Amazon region are deeply connected with the natural environment, land, and weather. Their daily routine is driven according to the flow of the river, whether is high or low tide, the weather, day, or night. Nature defines their culture, myths, and traditions (da Costa e Silva et al., 2010). Culture and tradition have a major influence on the riverine population's lifestyle and livelihood (Khouri, 2013; Lopes, 2013). Their fishery and hunting are influenced by their supernatural beliefs (Wagley, 1952; da Costa e Silva et al., 2010), which are related to spirits from the dead, and the changing of the seasons (da Costa Silva et al., 2010). The riverine population survived by subsistence economy, with family and community members (Khouri, 2013). Most of the

FIGURE 10.1 Image of a riverine house on the riverbank of the Xingu River.

Source: Fieldwork, 2018.

riverine families have lived away from cities, isolated inside the forest, most being illiterate, and having had low or no influence on decision-making and participatory engagement in public policies (da Costa Silva et al., 2010).

Barbosa & Carneiro (2017) explained that traditionally many of the riverine families from this region owned two houses, one on the riverbank (Figure 10.1) and another along the streams in the city center (Figure 10.2). This facilitated and allowed family members to stay longer periods in the city; this would depend on the family economic composition, health care needs, and if there are children over the age of ten. If the children graduate from the schools on the riverbank, it is common for the mother to move to the city for the children to continue their education. Families that sustain themselves through fishing would return to the city every weekend to sell their goods, different from the families that live off fruit pulp, eggs and chickens who go to the city every 15 days.

According to researchers, the sense of connection to the land and the community is strongly felt by riverine people (Khouri, 2013 and Costa Silva, no date), especially women (Peake, 2016). As Shiva (2016, p. 42) writes in her book chapter about the relationship between indigenous women and nature: "Women produce and reproduce life not merely biologically, but also through their social role in providing sustenance." In addition, "women's work is similarly invisible in providing sustenance and creating wealth for basic needs. Their work in the forest, the field

FIGURE 10.2 Riverine houses along the stream in the city center of Altamira were similar to these ones.

Source: Fieldwork, 2018.

and the river created sustenance in quiet but essential ways" (p. 44). In the case of this research, riverine women from that region of the Xingu River work together with their spouses in the fishery trade and farming. While the men catch the fish, the women clean it, and when men collect the crops, the women plant it. Selling the goods in the city's market is a way to increase the family's income. According to Parente (2015), many women lose their independence when resettlement takes place as this co-workmanship is eradicated.

With the construction of Belo Monte, riverine families were gradually resettled to five different Collective Urban Resettlement sites (Figure 10.3). The separation of neighbors and family members during the resettlement process started in 2014, which consequently created a social disarticulation and affected people psychologically (Villas-Boas et al., 2015). Being resettled to the outskirts of the city of Altamira, due to the reservoir formation, disconnected them from to the river, changing their traditions and disturbing what was known to them as their livelihood, consequently losing their identity and the sense of belonging (Barbosa and Carneiro, 2017; Villas-Boas et al., 2015).

Currently, their new lands are an urban plot of 300 square meters, extremely small for raising animals for their own consumption, growing, and farming their own food, resulting in no vegetable production as an income resource. The distance

FIGURE 10.3 Image of Collective Urban Resettlement Jatoba in Altamira.

Source: Fieldwork, 2018.

created between the resettlement sites to the river and the size of the plots has forced the local riverine people to buy food instead of growing it, catching it, or hunting for it. The lack of knowledge of riverine traditional livelihood by the authorities responsible for the mitigation program has an outcome that does not allow the riverine folk to excel and competently survive in such a location with their current knowledge and skills and abilities. This has had a negative impact on those affected and has resulted into an impoverished community who have lost their identity (Barbosa and Carneiro 2017).

10.2.4 Methodology

The fieldwork that was conducted was divided into two phases: the first fieldwork was conducted between March and May of 2018, and the second took place in January and February 2020. Eighteen riverine families were interviewed from four out of five of the Collective Urban Resettlement sites. Most of them had owned two houses prior to the resettlement, of which one house was situated in Altamira and one along the river. Only three interviewees did not own a previous urban house, the reason for that was because the previous riverbank house was close enough to the city to be able to utilize all facilities offered without having to possess a second house in Altamira.

The method used for this research was ethnographic; it is a method that allows one to understand a society from the inside, bringing alight the members' point of view (Gold, 1997) in a subjective and objective way (Hegelund, 2005); the researcher spent in total five months in the city of Altamira. Totally, 3 couples, 4 riverine men, and 11 women in their houses in the Collective Urban Resettlement sites were interviewed. The interviews conducted were in-depth open-ended multiple interviews that were able to explore the sequence from when the construction of the dam was announced, to the registration of every family, the negotiation about the compensation, resettlement and finally their current livelihood status; how their lives have changed with this shift of environment and how they have adapted to the new situation, urban space and houses. The participants who were willing to share their stories in their homes were chosen for the interviews. They were purposely selected in order to understand and to experience their living environment and analyze the entire context.

Twelve key expert interviews were also conducted with a carefully selected group of experts who were able to provide specific information regarding the city and the riverine community, and they were able to recall the whole experience from the perspective of the citizens in the development project. These key experts were professors from the local "Federal University of Para," one who was involved in traditional community studies and one was a social scientist who has written many articles on the impact of the dam to the riverine community; there were three representatives from a national NGO "Instituto Socioambiental" (ISA) who worked closely with the affected riverine communities, two experts from another NGO called "Movement of Those Affected by Dams," another expert from the NGO "Xingu for All" who works closely with the riverine people, two representatives from the local Women's Movement, the coordinator of a particular riverine area called the "Extractive Reserve," the local highest representative from the "Brazilian Institute of Environment and Renewable Natural Resources," and questions were answered via email to the "Government House" and "Norte Energia S.A."

A combination of the participatory walk-along method was also applied, where the researcher joined a boat trip with two riverine men to the Xingu River. During this trip, the interviewees shared a lot of valuable information: They pointed out the location of islands that were currently flooded, their previous fishing locations, the beaches that were used for camping during the fishing period, and where the riverine communities were located along the river. We also visited a riverine family home on the river and talked to the family. Afterward, the interviewer was joined an expedition that was organized by the "Instituto Socioambiental," for a week into the Amazonian Forest, sleeping in a riverine family's home alongside the river, spending the days with them, listening to their stories, and observing their routines. The interviews were recorded with the consent of the interviewees; every participant's name has been altered due to privacy reasons. Throughout the fieldwork, field notes were transcribed daily or recorded and every recorded interview was typed out, coded, compared, and analyzed. Another important source of information was the gathering of local articles and documents during the fieldwork that was used as additional information for the findings and literature review.

10.2.5 Findings

Traditionally, in that region of the Xingu River, riverine women and men work together in both the fishing industry and farming. Men and women have their own duties and is assigned due to physical strength and abilities; while one cleaned the house, the other repaired the fishing net; while the women made the meals, the men would make the juice or collect herbs in their plantation. Their children would also join the couple and, depending on their age, there would be different responsibilities. These responsibilities would vary, either by accompanying the parents during every task or the children would stay home doing the household chores, preparing the meals, and taking care of the younger children. In a riverine family, each member has their own tasks, and everyone relies on each other for the job to be done.

While living in the city for a period for the children's education, the women would either find a job in the city to add to the household income or occupy a spot in the local market to sell the product brought by the husband from the riverbank. With this income, the riverine women would generate their own income and not have to depend on the men.

Currently, after resettlement, the couples do not work together anymore; this whole setup has changed due to access to the river. This fact has increased the men's workforce and isolated women to the household focusing on childcare and household duties, weakening women's position in the household, and reinforcing Cernea's (2000) indicator of "impoverishment after resettlement."

The lack of a direct connection to the river and natural resources translates into the loss of traditional livelihood and has resulted in the loss of the riverine folk's identity. Ms. Flor said, "I will create a small forest on my allotment, I'll put chickens and plant trees."

Amazon Watch (2010) published a letter written by Xingu Alive Forever Movement of how important the river is to the people who have a close relationship with it:

> The river is the heart of our land and our people. Our lives are organized around this river. (…) It is the main path to our lands, our schools, our cemeteries, and our sacred sites. The river is our gate to the rest of the world. We (…) have a relationship of love and deep respect for the river, for its life and its people.

For a riverine, a home is not limited to the construction of the residence, but of what is found in the land that surrounds the house. Resettlement took away both land and house from them. Ms. Ana (interview, 2018) added, "life here [at the CUR] is very stressful, I can't go anywhere here, at least there [riverbank] I could go to the river (…) now the only time I have contact with water is when I take a shower." The psychological marginalization created after resettlement could be identified in their stories and complaints. RW9 is an example of many other riverine women who had their livelihood opportunity reduced and had to reorganize and relearn how to live in this new environment while having the knowledge and expertise of living in the natural environment. The lack of spatial identity and social elements that were needed to be included in the planning of the CURs left women vulnerable. Their livelihood assets as the services, work, education, and social networks have been dispersed, which have resulted into an "unadaptation" and the need to create an environment where they feel comfortable and recognizable. Throughout the first years of resettlement, riverine women tried to raise chickens in the small CUR allotment and plant various plants and trees. However, the lack of shade killed many of the small animals, and the poor soil was not strong enough for the plants and trees to survive.

Isolation and marginalization have become a reality for riverine women after resettlement. With the announcement of Belo Monte dam and the enormous number of migrants to work in the construction, the city became expensive and unsafe with high rates of violence (Marin and da Costa Oliveira, 2016). Due to the lack of preparation by the local authorities and federal authorities, the level of thefts in the CUR (Collective Urban Resettlement) became extremely high to the point that people would avoid the streets after sunset and most families placed steel grids to protect the windows and doors (Figure 10.4). Due to that new reality one member of the families that were interviewed was selected to guard the house while the other went out to work; these roles would always be assigned to the women. That fact not only affected women's livelihood, but also her connection to the family and community. Without being aware of it, riverine women grew deeper into the household responsibilities, lost their freedom and independence.

The lack of organization when distributing the riverine community to the different CURs fragmented the community members and families. Ms. Tania said, "my sister now lives in another CUR, it takes me over an hour to reach her house. Now I only see her once or twice a month. Whereas before we would see each other almost every day." This social disconnection has affected not only women's leisure and common use of space, but it has affected the family's food security as well. Riverine families have always relied on one another to survive; neighbors would help each other sharing their food but also helping around their land if someone was not able to do this. This trust is only created through a deep bond that was created over many years that was found in the riverine communities, which enabled their subsistence economy to work well between them.

A riverine woman whom I interviewed during my 2018 fieldwork said, "here we go hungry, there [forest] we could pick the fruits to eat at any time." With the reduction of income, social disconnection and being resettled away from their main food sources, food can become scarce and selective, which can in turn create the scenario of food insecurity (Barbosa & Carneiro 2017). Food insecurity became a

FIGURE 10.4 Image of a riverine woman in front of her house with windows and door protected by steel grids in one of the Collective Urban Resettlement.

Source: Fieldwork, 2018.

reality amongst the riverine community; the reduced and inaccessible land and distance from the river have disabled and discouraged them to use their knowledge. In addition, the environmental landscape has changed, which has created an impact in the natural resources found in nature. This fact has decreased the riverine family's income.

As a consequence of the feelings of injustice, the riverine community together with NGOs and a local legal prosecutor created the riverine council in 2018 (Mariuzzo, 2018). With this council, the riverine folk have a voice that can be heard and are now able to request their traditions and livelihood back. The council is composed of the impacted riverine people, and each community affected by the formation of the reservoir. Each community has their own representatives, and some of them are riverine women. Women who became depressed and vulnerable when resettled are now representing a part of their community and requesting their return to the riverbank to regain their traditions and their livelihood. The possibility of fighting for their rights and showing their indignity has given a voice to the women and has stimulated riverine women to learn how to read and write and to understand public policies. This community initiative has empowered women who are now being heard and sharing their thoughts to fight with and for their community's rights.

10.3 Conclusion

To conclude, this chapter has shown the loss in many layers of riverine women's livelihood and daily traditional habits after resettlement and disconnection to the river and forest. Answering the research question: How has the construction of Belo Monte dam in the Xingu River changed riverine women's livelihood traditions? Not having the natural environment to rely on as an income resource and leisure has changed the riverine women's daily routines. They feel that their belief system, knowledge, myths, and connection to nature is no longer relevant in the current living locations.

The main issues were the detachment of the people from their natural resources and the current insecure situation in the CURs that has created and deepened their dependence on their husbands and isolated them from their family and community. Overall, resettlement has had a negative impact on the riverine communities, who now have economic troubles and social disconnections. Poor project planning has fragmented the communities and has resulted in social disruption, which has marginalized the riverine people, especially the riverine women. The fact that the women are now the members of the family who stay at home instead of the men has proven that the patriarchal culture has been introduced into this society.

Impoverishment is a real consequence of resettlement, and it is experienced differently for each person; in addition to the shrinking of livelihood and space of leisure, women tend to be more isolated and have become more dependent on men. In the riverine tradition, women and men participated in different duties according to their physical capacity outside of the household duties, even though women were always the ones connected to the household responsibilities. Currently, due to the disconnection from the river, riverine women are constantly inside the house and are isolated from their social connections and daily external work, which has resulted in a high level of depression and vulnerability amongst them.

The voice and feelings of owning and having control of one's life were taken when the families were resettled to the outskirts of the city and away from the social connections and livelihood. Some of the riverine women have found a way within their new environment to overcome this feeling by being a part of the riverine council, which has strengthened and empowered these women. They now have the possibility to fight for their communities' rights and try to return to the riverbank, to their livelihoods and traditions.

It is important to include the issues of gender equality within the development process such as one of the SDG goals, as it will have a positive impact on the health and security of cities. It is important to review social customs and give the same opportunities to both men and women. Currently, there are more women living in poverty than men, seeing women as equal and giving the same opportunity and space as men can improve not only their own lives but everyone's around.

Note

1 IDMC. 2017, Dams and internal displacement, an introduction. www.internal-displacement.org.

References

Amazon Watch (2010). From the Xingu Alive Forever Movement: a letter of denouncement and indignation against the approval of the Belo Monte Hydroelectric Dam's provisional license < https://amazonwatch.org/news/2010/0204-from-the-xingu-alive-forever-movement-a-letter-of-denouncement-and-indignation-against-the-approval-of-the-belo-monte-hydroelectric-dams-provisional-license> Accessed: 15 December 2020

Anand, A. and Tiwari, G. (2006). A gendered perspective of the shelter–transport–livelihood link: The case of poor women in Delhi. *Transport Reviews*, 26 (1), pp. 63–80.

Asthana, R. (1996). Involuntary resettlement: Survey of international experience. *Economic and Political Weekly*, 31 (24), pp. 1468–1475. Available at: www.jstor.org/stable/4404268 [Accessed 14 May 2019].

Bajpai, D. and Gautam, U. (2018). Assessing social impacts of slum displacement on women's lives: A case study of three resettlement colonies in Delhi, India. *Kathmandu School of Law Review*, 6 (1), pp. 86–102.

Barbosa, M. and Carneiro, M. (2017). A expulsao de ribeirinhos em Belo Monte: relatorio da SBPC. Sao Paulo. Available: http://portal.sbpcnet.org.br/livro/belomonte.pdf. Accessed: 28 December 2020

Bisht, T. C. (2009). Development-induced displacement and women: The case of the Tehri Dam, India. *The Asia Pacific Journal of Anthropology*, 10 (4), pp. 301–317.

Cernea, M. M. (1997). Hydropower dams and social impacts: a sociological perspective. *The World Bank*.

Cernea, M. M. (2000). Risks, safeguards and reconstruction: A model for population displacement and resettlement. *Economic and Political Weekly*. Available at: www.jstor.org/stable/4409836.

Cernea, M. M. (2006). Re-examining "displacement": a redefinition of concepts in development and conservation policies. *Social Change*, 36 (1), pp. 8–32.

Cernea, M. M. (2008). Reforming the foundations of involuntary resettlement: Introduction. In: M. M. Cernea and H. M. Mathur (Eds.), *Can compensation prevent impoverishment? Reforming resettlement through investments and benefit sharing*. New Delhi: Oxford University Press, pp. 1–8.

Cernea, M. M. (2009). Financing for development: Benefit-sharing mechanisms in population resettlement. In: A. Oliver-Smith (Ed.), *Development and dispossession: The crisis of forced displacement and resettlement*. Santa Fe: School for Advanced Research Press. pp. 1033–1046.

Cernea, M, M. and Maldonado, J. (2018). *Challenging the prevailing paradigm of displacement and replacement*. New York: Routledge.

Chaves, K. A., Monzoni, M., and Artuso, L. F. (2019). UHE Belo Monte: Reassentamentos rurais, participação social e direito à moradia adequada. *Revista Direito GV*, V. *15* N.2.

Coelho, K., Venkat, T., and Chandrika, R. (2013). Housing, homes and domestic work, a study of paid domestic workers from a resettlement colony in Chennai. *Economic and Political Weekly*, 48 (43), pp. 39–46. Available at: www.epw.in/system/files/pdf/2013_48/43/Housing_Homes_and_Domestic_Work.pdf [Accessed 13 April 2013].

Courtland Robinson, W. (2003). Risks and rights: The causes, consequences, and challenges of development-induced displacement, an occasional paper. *The Brookings Institution – SAIS Project on Internal Displacement.*

da Costa Silva, J. (no date). O mito e as crencas como constituintes do espaco ribeirinho na formacao do modo de vida amazonico. Available at: www.neer.com.br/anais/NEER-1/mesas/josu%C3%A9.pdf

da Costa Silva, S. S., Pontes, F. A. R., dos Santos, T. M., Maluschke, J. B., Mendes, L. S. A., dos Reis, D. C., and da Silva, S. D. B. (2010). Rotinas familiares de ribeirinhos amazônicos: uma possibilidade de investigação. *Psicologia: teoria e pesquisa*, 26 (2), pp. 341–350.

de Wet, C. J. (ed). (2006). *Development-induced displacement: Problems, policies and people* (vol. 18). New York: Berghann Books.

Energia, N. (2011). Projeto Básico Ambiental. Plano de Atendimento a Populacao Atingida. *Versão Final*, 1. Available at: www.mdpi.com/2071-1050/9/8/1422/pdf [Accessed 28 May 2019].

Gold, R. L. (1997). The ethnographic method in sociology. *Qualitative Inquiry*, 3 (4), pp. 388–402.

Hall, A. and Branford, S. (2012). Development, dams and dilma: The saga of Belo Monte. *Critical Sociology*, 38 (6), pp. 851–862.

Hegelund, A. (2005). Objectivity and subjectivity in the ethnographic method. *Qualitative Health Research*, 15 (5), pp. 647–668.

Instituto Igarape. (2018). *Observatório de Migrações Forçadas.* Available at: https://igarape.org.br/wp-content/uploads/2018/03/2018-03-06-factsheet-migracoes.pdf. [Accessed 20 March 2021].

Khouri, H. (2013). Involuntary Displacement and the Belo Monte Dam: Changes in Self-Perception of the Volta Grande´s Displacement Riverine Peasants (Doctoral dissertation) Mount Holyoke College.

Koenig, D. (2009). Urban relocation and resettlement: Distinctive problems, distinctive opportunities. In: A. Oliver-Smith (Ed.), *Development and dispossession: The crisis of forced displacement and resettlement.* Santa Fe: School for Advanced Research Press, pp. 119–139.

Lopes, S. (2013). Povos e comunidades tradicionais direitos humanos e meio ambiente. *Lex Humana*, 5 (1), p. 160.

Marin, R. E. A. and da Costa Oliveira, A. (2016). Violence and public health in the Altamira region: The construction of the Belo Monte hydroelectric plant. *Regions and Cohesion*, 6 (1), pp. 116–134.

Mariuzzo, P. (2018). Ribeirinhos de Belo Monte. *Ciência e Cultura*, 70 (1), pp. 6–8.

Mathur, H.M., (2016). *Assessing the Social Impact of Development Projects: Experience in India and Other Asian Countries.* Geneva, Switzerland: Springer International Publishing.

McDonald-Wilmsen, B. and Webber, M. (2010). Dams and displacement: Raising the standards and broadening the research agenda. *Water Alternatives*, 3 (2), pp. 142–161.

Mehta, L. (2009). The double bind: A gender analysis of forced displacement and resettlement. *Displaced by development: Confronting marginalisation and gender injustice*, Mehta, L. (ed.) pp. 3–33. New Delhi: Sage.

Mehta, L. (2011). The settler and his wife, gender and the politics of displacement. *Rivista Telematica Di Studi Sulla Memoria Femminile*, 17, pp. 25–43.

Movimento dos Atingidos por Barragens (2004). Ditadura contra as populações atingidas por barragens aumenta a pobreza do povo brasileiro. Brasília, 2004. Website link: www.alainet.org/pt/active/5807. [Accessed: 20 March 2021]. [Links]

Muggah, R. (2008). *Relocation failures in Sri Lanka: A short history of internal displacement and resettlement*. London: Zed Books.

Nguyen, H. T., Pham, T. H., and Lobry de Bruyn, L. (2017). Impact of hydroelectric dam development and resettlement on the natural and social capital of rural livelihoods in Bo Hon Village in Central Vietnam. *Sustainability*, 9 (8), p. 1422.

Oliver-Smith, A. (2006). Displacement, resistance, and the critique of development: From the grassroots to the global. In: C. De Wet (Ed.), *Development-induced displacement: Problems, policies and people*. Oxford and New York: Berghan Books, pp. 141–179.

Papmehl-Dufay, L. (2015). Places that matter. *Landscape and Heritage Studies*, 143, pp. 143–161

Parente, T. G. (2015). Gênero e (in) sustentabilidade de mulheres nos reassentamentos rurais da Usina Hidrelétrica de Estreito–Tocantins. *OPSIS*, 15 (2), pp. 399–416.

Peake, L. (2016). The twenty-first-century quest for feminism and the global urban. *International Journal of Urban and Regional Research*, 40 (1), pp. 219–227.

Sahoo, D. and Mishra, N. (2016). Development-induced displacement and gender injustice: Some critical feflections. *Journal of Politics and Governance*, 5 (4), pp. 19–30.

Samling, C. L., Ghosh, A. K., and Hazra, S. (2015). Resettlement and rehabilitation: Indian scenario. Working Paper, Deltas, Vulnerability and Climate Change: Migration and Adaptation, IDRC Project Number 107642. Kolkata: DECCMA. Available at: http://hdl.handle.net/10625/57387 [Accessed 14 May 2019].

Shiva, V. (2016). *Staying alive: Women, ecology, and development*. New Delhi: Kali for Women.

Sikka, G. and Mathur, V. (2018). Gender analysis of post resettlement transformation in livelihood opportunities at Sardar Sarovar Project resettlement sites, India. *Journal of Land and Rural Studies*, 6 (1), pp. 1–14.

Smyth, E., Steyn, M., Esteves, A. M., Frank, D. M., & Vaz, K. (2015). Five 'big' issues for land access, resettlement and livelihood restoration practice: findings of an international symposium. *Impact Assessment and Project Appraisal*, 33 (3), pp. 220–225.

Terminski, B. (2013). Development-induced displacement and resettlement: Theoretical frameworks and current challenges, Geneva, May 2013, p. 177. Available at: https://dlc.dlib.indiana.edu/dlc/bitstream/handle/10535/8833/Bogumil%20Terminski,%20developmentinduced%20Displacement%20and%20Resettlement.%20Theoretical%20frameworks%20and%20current%20challenges.pdf

Terminski, B. (2015). *Development-Induced Displacement and Resettlement: Causes, Consequences and Socio Legal Context*. Stuttgart: ibidem-Verlag.

Van Eerd, M. (2018). Development with impoverishment: Displacement, resettlement and inadequate compensation in India and Nigeria. In: B. Hoops, E. J. Marais, L. van Mostert, Schalkwijk, and N. K. Tagliarino (Eds.), *Rethinking Expropriation Law III, Fair Compensation*. The Hague: Eleven International Publishing, pp. 87–119.

Van Eerd, M. (2017). Pushed to the bush? Changes in resettlement approaches and its impact on the urban poor in India. In: M. P van Dijk, J. Edelenbos & K. van Rooijen (Eds.), *Unfolding city governance in complex environments*. Rugby: Practical Action.

Van Eerd, M. (2008). *Local initiatives in relocation: The state and NGOs as partners? From research to action*. New Delhi: Ajay Kumar Jain for Manohar Publishers and Distributors.

Villas-Boas, A., Garzon, B. R., Reis, C., Amorim, L., and Leite, L. (2015). Dossie Belo Monte: Não há condições para a licença de operação. *Instituto Socioambiental (ISA), Brasilia, DF, Brazil.*

Wagley, C. (1952). Race relations in an Amazon community. *Race and class in rural Brazil.* Paris: Unesco, pp. 116–141.

World Bank. (2004). *Involuntary Resettlement Sources Book: Planning and Implementation in Development Projects*. Washington, DC: The International Bank for Reconstruction and Development.

World Bank. (2015). Resettlement Fact Sheet. Washington, DC: World Bank Group.

Zaman, M, Nair, R., and Guoqing, S. (2022). *Resettlement in Asian countries. Legislation, administration and struggles for rights*. Oxon: Routledge.

11

REIMAGINING THE GENDER ROLES WITHIN JAPAN TO ACHIEVE SUSTAINABLE GENDER EQUALITY

Sarah Parsons

11.1 Introduction

Japan's rapidly ageing population and low birth rate have led to predictions that 12 million Japanese people may disappear from the country's workforce by 2040 (Fleming, 2019) and have already resulted in labour shortages in some industries (Jones and Seitani, 2019). In order to boost the female labour market and the birth rate, the Japanese Government introduced their 'Womenomics' initiatives in 2013 (Matsui et al., 2014). These included the introduction of family-friendly policies for working families such as more childcare provision and generous parental leave benefits. One of the successes of the Womenomics initiatives was an increase in the female labour participation rate to a record high of 71% in 2019 (Matsui et al. 2019) although the majority of this rise happened in non-regular roles, which was ultimately unsustainable. Many of the women within these non-regular roles have borne the brunt of Covid-19 job cuts (Zhou, 2021). Although 'continuing to expand female employment must remain a top priority for the Japanese government and society' (Matsui et al., 2019), these initiatives failed to achieve a sustainable inclusion of women in the Japanese workforce, and Japan has made little progress in achieving the key targets and indicators of the UN's Sustainable Development Goal (SDG) 5: achieve gender equality and empower all women and girls (UN, no date). The sustainable inclusion of women into 'all aspects of social, economic, political, and academic activities' (Kano, 2015: 100) by achieving the key SDG targets will be referred to throughout this chapter as 'sustainable gender equality'.

I believe that the entrenched cultural norms in Japan have created insurmountable barriers to sustainable gender equality and have underpinned Japan's failure to deliver on the SDG targets. This chapter will first look at why Japan has not been able to meet the SDG targets. It will then briefly look at why sustainable gender equality is so important for Japan and will explore the cultural norms with

DOI: 10.4324/9781003174257-14

examples of current case studies that highlight the barriers these norms cause. It will finally introduce a potential way for Japan to eliminate these barriers by reimagining the gender roles from a grassroots level in a culturally appropriate way. Although these norms will be discussed within a gender binary context, the 'gender review' suggested within the final section offers scope for inclusion of other gender identities that are not currently included within the discourse of equality, diversity, and inclusion in Japan.

11.2 Sustainable gender equality

11.2.1 Failure to achieve the SDGs

Since the implementation of the Womenomics initiatives, Japan has scored consistently low on the rankings of the World Economic Health Forum's Global Gender Gap Report. Japan currently ranks 120th out of 156 countries, one of the lowest rankings amongst G7 countries, with particularly low rankings on the subsections of economic participation and opportunity and political empowerment (WEF, 2021). The initial Womenomics targets aimed at getting 30% of leadership positions occupied by women by the end of 2020 were subsequently downgraded and have recently been postponed for ten more years after it became clear that they were unachievable (Parsons, 2020). These initiatives failed to achieve SDG Target 5.5: improve women's full and effective participation and equal opportunities for leadership at all levels of decision-making in political, economic, and public life.

The Japanese employment system has been based on a male-breadwinner female-dependent model which has resulted in a gendered employment system and high levels of gender segregation within the workplace (Macnaughtan, 2015; Takami, 2018) as well as high levels of gender bias and institutionalised discrimination, examples of which will be discussed further in this chapter. This failure to tackle the high levels of gender bias and discrimination against women in the workplace represents a failure to reach SDG Target 5.1.

Historically, the female labour force has been used as a buffer rather than a sustainable labour source and women are overly represented within non-regular jobs (Macnaughtan, 2015; Kano, 2018). The division of labour is embedded in social systems which place a high dependency of unpaid work on women (Macnaughtan, 2015; Shirahase, 2015; Takami, 2018). Men still face social stigmas around their involvement in domestic work, and many Japanese women face the choice of families or full-time work (Brinton and Oh, 2019; Yoshikawa et al., 2018). Japanese women are still held back by the gendered division of labour and do most of the unpaid work. This represents a failure to achieve SDG target 5.4: recognise and value unpaid care and domestic work through the provision of public services, infrastructure, and social protection policies and the promotion of shared responsibility within the household and the family as nationally appropriate.

11.2.2 Sustainable gender equality imperative

The world's third largest economy cannot be sustainable when half of the population are acting as a buffer labour market and facing cultural and structural barriers to inclusion and leadership. Sustainable gender equality would also result in more sustainable globalisation activities for Japanese firms including better talent retention and a raised global reputation. There is currently an increased focus from international asset managers on whether companies are meeting their SDG goals (Matsui et al., 2019), and major global investors have started putting pressure on Japanese companies by threatening to vote against boards of directors at Japanese companies that have low levels of gender board diversity (Taylor, 2021). Sustainable gender equality is therefore an important imperative for Japan.

This chapter avoids linking the fertility rate to sustainability for several reasons: (1) sustainable gender equality in Japan should not be pursued just to raise the birth rate; otherwise, this perpetuates the emphasis of the role of a mother for Japanese women, (2) allowing women to move out of low-paid non-regular employment into more secure employment with a better division of domestic labour, better cultural acceptance of working mothers, and a less gendered workplace may naturally boost the birth rate (Osawa and Kingston, 2015; Shirahase, 2015), and (3) sustainable gender equality should still be pursued even if it does not lead to a higher birth rate (Kano, 2015: 100).

11.3 Gender norms in Japan

Gender norms within this chapter will be defined as 'social norms defining acceptable and appropriate actions for women and men in a given group or society, embedded in formal and informal institutions' (Cislaghi and Heise, 2020: 415–416). Although norms are only one element of the gender system, they represent the social rules and expectations that keep the gender system intact. It is important to understand how these norms have led to the socialisation of certain roles for Japanese women and how these roles negatively impact upon sustainable gender equality.

11.3.1 Roles of a Japanese woman

Japan is a deeply patriarchal society influenced heavily by the Confucian concept of filial piety. The concept of the household head is still defined in patriarchal terms, and according to the 'Kosei' (family registration system), a Japanese wife is defined as the property of her husband's household (Sugimoto, 2010: 158). The Japanese Government only recently overturned a law stating that Japanese women could not marry for six months after a divorce but have yet to eliminate the male-only right to deny legal fatherhood. These patriarchal and Confucian influences combined with specific economic conditions have resulted in sharply divided gender roles in Japan.

11.3.1.1 Woman as mother/wife/educator

The role of mother/wife is one of great historical and cultural importance in Japan.

The ideology of 'Good Wife, Wise Mother' used within female education in the Meiji era perpetuated the Confucian emphasis of a wife's obedience to her husband and tied this role to the good of the nation (Usui et al., 2003). The post-war economic miracle in Japan benefited from the wives of the well-paid and hardworking salary men fulfilling the role of the housewife by managing all the household affairs whilst the men worked long hours. The long hours were exacerbated by obligatory after-hour activities to strengthen workplace bonds and show commitment to a 'Japanese work ethic' that was based on loyalty to the company and relationship-based business practices (Mukherjee, 2015). These housewives played a 'pivotal role in marshalling the savings that financed the economic miracle' (Murphy, 2014: 159) and the 'backstage work' was seen as a housewife's 'duty' to endure to make sure the husband could fulfil his role as breadwinner (Condon, 1985). The idealised view of this role helped to embed a social construction of motherhood (Mukherjee, 2015). Mothers were socialised into becoming 'good mothers' and were expected to be the primary socialisation agent for their children within the first three years of their life with high levels of involvement in their offspring's education. Working mothers were subjected to a process of social control whereby they were expected to prove they were good mothers by providing a high level of care for their children and taking on the full responsibility for childcare arrangements. This often resulted in criticisms of working women who were not dedicating themselves fully to their child's development and led to many women opting out of the workforce due to the unrealistic demands from both these roles (Fujita, 1989). Although many women have since gone back into the labour market, the expectations of the maternal role within their children's upbringing are still high due to demands placed on mothers by schools, societal expectations, and peer pressure (Kano, 2016).

The 'male-breadwinner and female-caregiver' model fitted an ideological purpose within Japan's post-war economic growth, when lifetime employment was the norm in the larger companies and long working hours and total dedication to the company were expected of white-collar male workers. Gender segregation within the workplace was normalised through the introduction of dual career paths for men and women (Macnaughtan, 2015), which has contributed to the peripheral role women have played within the labour market (Benson et al., 2007) and their role as a 'buffer' non-regular source of labour when needed (Houseman and Abraham, 1993). This segregation was emboldened by strong societal beliefs that married women who worked full time should leave the company either when they married or certainly when they were pregnant. These core beliefs still hold considerable influence within current Japanese society and corporate practices, and they have real-life implications for women who challenge these.

11.3.2 Effects of these norms

These conservative attitudes towards the role of a woman as caregiver/mother/wife are firmly entrenched within the core seats of power in Japan, as evidenced by recent comments by leading Japanese politicians promoting the primary role of Japanese women as 'baby makers' (McCurry, 2015). One male politician even went as far as to admonish a female politician in Parliament telling her to 'go and have babies' (BBC, 2014). The following examples show how these attitudes have been used to discriminate against and exclude women from the workplace and villainise those who challenge the status quo.

- **Incompatibility of working life and childcare responsibilities**
  In 2017, a female Kumamoto city assembly member who brought her newborn baby into the chamber to breastfeed was removed and forced to apologise. New regulations were subsequently drafted prohibiting anyone other than adults to enter the room. The member alleges that she has since been subjected to further criticisms of her abilities as a mother and other council members have expressed the view that 'it's outrageous to provide day care services for council members with young children' (McCurry, 2018). Not only does this highlight an entrenched discriminatory attitude towards working mothers in Japan but also highlights how norms and attitudes can be either barriers or facilitators to female participation in politics.
- **Maternity Harassment (Matahara) in Japan**
  According to the Japan Institute for Labor Policy and Training's Survey on Unfavourable Treatments Because of Pregnancy and Sexual Harassment in 2015, the rates of maternity harassment victimisation were 22.3% for regular workers and 15.4% for non-regular workers (JILPT, 2016). A website that advocates for victims of Maternity Harassment in Japan (Mataharanet) reports that this harassment sometimes takes the form of an imposition of traditional gender values. Some working mothers have been harassed if they do not want to leave the labour force after becoming pregnant/having children, some have had promotional opportunities taken away from them since it is seen as being 'too hard for them', and some have been made to feel guilty for adding on the burdens of extra work through their 'selfish' attitudes if they want to take maternity leave or need more flexible working hours. The entrenched norms and expectations that women should prioritise their role as a mother result in discrimination and barriers to sustainable gender equality.
- **Expectations of the maternal role and childcare**
  The Japanese Consulate in Edinburgh was fined for discriminating against a Japanese woman within an interview by asking her about her childcare arrangements (Learmonth, 2016). Despite the fact that this is an illegal practice in the UK, the entrenched cultural expectations that women should be responsible for childcare resulted in this questioning of her ability to do a job whilst fulfilling the role of a mother.

- **Women of childbearing age are not productive workers**
  Tokyo Medical University made the news in 2018 (BBC, 2018) after they were found to have been manipulating exam scores by setting a lower pass mark for men than for women to ensure that enough male graduates entered the medical profession. This had been an institutionalised practice for many years, and the same practices were discovered in other medical universities in Japan. One of the explanations given for this practice was that female doctors tend to quit the profession when they start families, which creates problems such as staff shortages. The imposition of traditional gender roles results in discrimination and prevents sustainable gender equality.

11.3.2.1 Double burden-unpaid labour

Another key barrier to sustainable gender equality is the disproportionate double burden of unpaid care work that falls on Japanese women due to the entrenched male-breadwinner female-dependent model. According to the latest available OECD statistics, Japanese men spend the least amount of time participating in unpaid work (40.8 minutes a day compared to an OECD average of 136.5 minutes) and the most amount of time in paid work (451.8 minutes a day compared to an OECD average of 317.8 minutes) amongst all the OECD countries plus China, South Africa, and India (OECD. Stat, no date).

11.3.2.2 Paternity leave

Although the Japanese Government offers one of the most generous parental leave packages in the world, only around 3% of Japanese fathers take paternity leave (Broom, 2020). This is such an unusual occurrence that when an up-and-coming politician recently announced he was going to take paternity leave, it caused a media storm. Two prominent paternity court cases made headlines in 2019, when the men involved claimed they were subject to demotions and harassment due to their decision to take this leave (Rich, 2019). The cultural norms that stigmatise Japanese men who want to move out of the traditional norms perpetuate the double burden for women.

11.4 Behavioural expectations of Japanese women

I will now look more closely at the archetypal feminine roles available to Japanese women-archetypes that influence and shape certain expectations of behaviour.

11.4.1 Nadeshiko

Nadeshiko, an abbreviation of Yamato nadeshiko, is a phrase used to describe an ideal Japanese woman. It was originally attributed to noble women in the 19th century who were operating within the patriarchal behavioural requirements of that time, which included feudal loyalty and filial piety. A Nadeshiko woman had a strong spirit but was expected to have a quiet mental strength, due to the lack of voice that women had in the public arena. Over the years, this label has been used to refer to women who embody features considered ideal for Japanese women such as loyalty, modesty, domestic ability, gentleness, gracefulness along with self-discipline. Such is the importance of the Nadeshiko archetype in representing an idealised image of Japanese women that the Ministry of Economy, Trade and Industry and the Tokyo Stock Exchange granted companies who had performed well in encouraging women's success in the workplace between 2012 and 2014 the 'Nadeshiko Brand' (JapanGov, no date). It is also the name for the Japanese national female football (soccer) team.

11.4.2 Kawaii

Kawaii is a popular cultural aesthetic originating in Japan that has also become an important cultural export for brands such as Hello Kitty and Doraemon. Definitions and representations vary, but it is normally used to describe purely innocent and immature things, anything that softens the heart (Ozaki and Johnson, 2013) or anything lovable and pretty. From the 1980s, the kawaii image became very popular within manga showing sweet women with large innocent eyes, and it has since developed into a cutesy childlike image. It is often used to describe the appearance and behaviour of girls and women who are kind, lovable, and non-aggressive with high-pitched voices. The extreme characteristics of this archetype are currently displayed by popular female pop stars such as Kyary Pamyu Pamyu and the Japanese idol band AKB48.

The idealised traits from both these archetypes (attractive, graceful, gentle, compliant, and pleasant company) are socialised from an early age in Japan both within language and behaviour. The use of specific language that gives more dominant words to men and more agreeable words to women is instilled into children's vocabularies from an early age (Davies and Kasama, 2004: 124). The term 'kawaii' has also been used within preschools to praise and socialise the expected behaviour of girls within this frame of 'cuteness' and 'compliance' (Burdelski and Mitsuhashi, 2010). Yamamoto (2016) found that mothers tend to value characteristics such as cuteness and gentleness over academic achievement and intelligence in their daughters, the latter being seen as a hindrance to her development due to concerns that 'smart' girls may be perceived as less 'cute' and therefore disadvantaged in the marriage market. A lack of social acceptance of the 'smart' archetype for Japanese women may account for the gendered differences within graduate paths in Japan. Women are still underrepresented in STEM jobs (Yoshikawa et al., 2018) and make up around only 20% of the student body at Japan's most elite university, the University of Tokyo, more commonly known as Todai (Kakuchi, 2020).

Although some of the aforementioned archetypal behavioural traits are not inherently negative, the deep socialisation of these as idealised feminine behaviour in many areas of Japanese society has resulted in negative consequences for some women who do not conform to this archetypal behaviour, even when some of the non-normative behaviour is needed for sustainable gender equality (e.g., speaking out, delegating, decision-making, leading meetings).

11.4.2.1 Speaking out

After more women were promoted to take part in Olympic Committee meetings, the Head of the Tokyo Olympic Committee (and former Japanese Prime Minister) complained that female behaviour made meetings last too long (Harding, 2021). Seen within a context of a cultural preference for subservience and compliant female behaviour, this is perhaps not an unusual reaction from older men in Japan, who are not accustomed to giving women a voice within decision-making processes outside of the home. However, when this attitude is institutionalised within workplaces, it acts as a major barrier to sustainable gender diversity.

11.4.2.2 Looking cute/feminine

In 2019, two prominent social media movements protested Japanese company dress codes and norms that mandate the wearing of high heels in certain roles and ban the wearing of glasses in others as they make the wearers look cold, do not suit traditional kimonos, and do not allow for their make-up to be seen (Steger, 2019). Such expectations of a feminine and attractive appearance for women in the workplace go beyond practicality and reasonable respectful dress standards.

In 2016, a well-known Japanese travel company launched a PR campaign competition for male executive travellers to win a chance to sit next to an attractive female student from Todai (Samadi, 2016).

Sustainable gender equality is unachievable whilst women are restricted within these roles and whilst their main value is their ability to be cute, non-confrontational, and pleasing company. These deeply entrenched beliefs about female appearances and behaviour have become firmly embedded within corporate practices. Although some Japanese women reject the normative gender roles and are reaching positions of leadership, all Japanese women are affected by the societal expectations placed upon them to get married and have children whilst taking on the double burden and adopting certain supportive feminine behaviours and appearances at work (Belarmino and Roberts, 2019). The narrow and entrenched gender roles underpinning these societal expectations are one of the key reasons that the 'womenomics' initiatives did not work as they did not challenge the narratives within these norms and instead placed unrealistic burdens on women who were encouraged to go back to work (Macnaughtan, 2015; Kano, 2018) and face a workplace full of gender bias and discriminatory barriers. Firm-level changes and structural support have a key role to play but will be no more than 'sticking plasters' over these cultural barriers. Japan must move away from the male breadwinner model and change the social norms surrounding gender roles to become 'a society that supports both traditional and alternative forms of families and pushes for women's increased participation in the labor force and decision making, as well as for male participation in housework and childcare' (Kano, 2015: 99). My view is that the only way to achieve this is through a reimagination of the gender roles within Japan.

11.5 How can the cultural norms and policies surrounding gender be reimagined in a culturally appropriate way for sustainable and meaningful change?

My recommendation is twofold. First, a review that would encompass a re-evaluation and reimagination of the gender roles followed by an implementation of the findings from this review at a grassroots level. This way of tackling gender bias and norms has been recently advocated by the Singaporean Government, who have recently launched an initiative called 'Conversations on Singapore Women's Development'. This is part of a wide review of the issues related to gender inequality involving partners from the public and private sectors, grassroots, and youth organisations to implement a mindset change from an early age (Government of Singapore, 2021).

11.5.1 Review-based approach

This review-based approach would work in Japan too, although there will be challenges involved in untangling and decoding the cultural influences that also shape gender inequality in Japan, such as 'keeping the harmony' and the Japanese concept of 'honne' (one's real truth) and 'tatemae' (the public face to avoid upsetting the harmony). These cultural norms hinder an open and frank sharing of honest opinions, especially when these opinions may go against the status quo. However, the tradition of written information-sharing is deeply embedded within Japanese organisations and parts of this review could also be done through a mixture of public lectures and consultations discussing the gender roles with opportunities for written feedback.

Careful attention must be paid to the selection of the parties/players within this review so that they do not just reflect the current status quo of opinion and institutionalised power in Japan. More and more people in Japan are speaking out against blatant examples of sexism, showing the desire of a certain segment of the Japanese population to change things. These voices, alongside those of well-respected Japanese feminist scholars, must be included. There is also a role for global best practice within this review although this may become problematic, especially if it presupposes the transfer of Westernised concepts of female empowerment or gender policies that currently have no cultural fit with Japan. This chapter has already highlighted the shortcomings of Womenomics in using standardised targets to achieve gender sustainability but there is an increasing amount of business literature recommending Western 'pipeline' management strategies to mobilise Japanese women into leadership, such as mentoring, sponsorship, target setting, and implementation as well as enhancing training (Matsui et al.,

2019; McKinsey, 2021). These strategies certainly have a role to play but will only be truly effective once there has been an acceptance of the cultural shift. I would also question whether these strategies have been effective in countries such as the UK, where progressive gender policies and support structures have been put into work environments where masculine traits of leadership are still recognised and rewarded over feminine traits (Usui et al., 2003; Parsons, 2017). This 'review' would give Japan the unique chance to forge their own 'gender' trajectory and explore the value of both masculine and feminine leadership traits to enact deeper institutionalised changes and inclusivity. There are certainly some aspects of the Japanese feminine archetype that do not necessarily need to be discarded and the current pandemic offers Japan new opportunities to explore and embed different working styles. There are also opportunities to reimagine the gender norms from a wider perspective and potentially bring in wider discussions involving non-binary definitions of gender.

11.5.2 Reintroducing the roles

The most appropriate place to reintroduce the reimagined norms would be in early years education, specifically in Japanese preschools. Not only does this align with gender schema theories, such as Bem's (1994) theory that children learn about male and female roles from the culture in which they live and adjust their behaviour to align with the gender norms of their culture from the earliest stages of social development, it also aligns with the role that Japanese preschools play within the socialisation of behaviours and cultural norms.

11.5.2.1 Japanese preschools-socialisation

Most Japanese children enter a preschool environment around the age of three before they start compulsory education at the age of six. Japanese preschools play a vital role in a child's social development and function as environments where children are socialised into the rules of operating within Japanese society by receiving a moral education that shapes their understanding of the norms and values of Japanese culture (Street, 2013). They learn the skills and behaviours needed to function within the group-oriented nature of Japanese society and understand their place in the hierarchies (Davies and Kasama, 2004; Burdelski, 2010).

Research on how gender is socialised in Japanese preschools shows how Japanese preschool teachers may position children in various interactional roles by channelling gender in everyday interactions. This research also shows observable differences within gendered behaviours such as girls speaking softly without force, feminine speech patterns impeding girls' participation in some communications, girls' lack of participation at storytelling time, and boys being more involved in forming hierarchies that value dominant males (Davies and Kasama, 2004; Burdelski and Mitsuhashi, 2010).

11.5.2.2 Implementation

The implementation of the review findings into an early years educational setting would be a complex task, not least because there does not seem to be a standardised global approach or method for integrating gender-related issues in preschool practices, due to the multi-dimensional and complex aspects of gender relations and the differing policies and cultures (Heikkilä, 2020). There are many complex issues involved, such as: who is responsible for implementing the gender framework and how that framework fits within wider societal contexts? It is clear that much more research is needed in this area. There is also more scope to further examine the gendered experiences of Japanese preschool children including the language used, the behaviour rewarded, the expectations of the relationship between school and mother, and the characters and situations the children encounter in stories. In 2016, I visited Stockholm University to give a lecture about the gender roles in Japan and asked the Swedish students about the feminine role models they had internalised as a child. The name they all mentioned was Pippi Longstocking, a popular book character of a nine-year-old girl who subverts the traditional gender roles by rejecting the expectations of certain behaviours and appearance that can so often hold women back. There could be opportunities for similar Japanese archetypal feminine behaviours and roles.

11.6 Conclusion

I have argued that a reimagination of gender roles is a viable way to achieve sustainable gender equality in Japan, and the suggestions I have given on how to do this will hopefully form the start of research on the different areas highlighted. These suggestions also offer Japan a unique advantage in that it can write its own 'gender' trajectory and hopefully include other representations of gender and diversity that have so far not been widely accepted within Japanese society. The implementation and realisation of such changes will take a generation to become socialised, so it is important that the process starts now. Although there is a possibility that generational attitudes may naturally change to achieve sustainable gender equality, the current trends within Japan suggest that younger people are rejecting the traditional male breadwinner female dependent roles but, with no viable alternatives, are moving increasingly towards the periphery of the labour market (Song, 2018) and rejecting real-life relationships. A healthy reimagination of these roles resulting in sustainable gender equality will benefit them and the future of Japan.

References

BBC News. (2014). Japan lawmaker admits to sexist remark. *BBC News*. Available at: www.bbc.co.uk/news/world-asia-27971062 (Accessed 23 April 2021).

BBC News. (2018). Japan medical schools 'rigged women's results.' *BBC News*. Available at: www.bbc.co.uk/news/world-asia-46568975 (Accessed 23 April 2021).

Belarmino, M., and Roberts, M. (2019). 'Japanese Gender Role Expectations and Attitudes: A Qualitative Analysis of Gender Inequality'. *Journal of International Women's Studies*, 20, pp. 272–288.

Bem, S. (1994). *Lenses of Gender: Transforming the Debate on Sexual Inequality*. New Haven, CT: Yale University Press.

Benson, J., Yuasa, M., and Debroux, P. (2007). 'The Prospect for Gender Diversity in Japanese Employment'. *International Journal of Human Resource Management*, 18 (5), pp. 890–907.

Brinton, M., and Oh, E. (2019). 'Babies, Work, or Both? Highly Educated Women's Employment and Fertility in East Asia'. *American Journal of Sociology*, 125 (1), pp. 105–140 https://doi.org/10.1086/704369

Broom, D. (2020). 'The 'future' prime minister of this country just made history by challenging gender stereotypes', *Weforum*, 7 Feb., Available at: www.weforum.org/agenda/2020/02/paternity-leave-japan-government-minister/#:~:text=Japan%20has%20among%20the%20most,fathers%20take%20any%20paternity%20leave (Accessed 27 April 2021).

Burdelski, M. (2010). 'Socialising Politeness Routines: Action, Other-orientation and Embodiment in a Japanese Preschool'. *Journal of Pragmatics*, 42 (6), pp 1606–1621.

Burdelski, M., and Mitsuhashi, K. (2010). 'She Thinks You're Kawaii: Socializing Affect, Gender, and Relationships in a Japanese Preschool'. *Language in Society*, 39 (1), pp. 65–93. Available online: www.jstor.org/stable/20622704. (Accessed 27 April 2021).

Cislaghi, B., and Heise, L. (2020). 'Gender Norms and Social Norms: Differences, Similarities and Why They Matter in Prevention Science'. *Sociology of Health & Illness*, 42 (2), pp. 407–422.

Condon, J. (1985). *A Half Step Behind: Japanese Women of the '80s*. New York: Dodd Mead & Co.

Davies, B., and Kasama, H. (2004). *Gender in Japanese Preschools: Frogs and Snails and Feminist Tales in Japan*. Cresskill, N.J.: Hampton Press.

Fleming, S. (2019). 'Japan's workforce will be 20% smaller by 2040', *Weforum*, Available at: www.weforum.org/agenda/2019/02/japan-s-workforce-will-shrink-20-by-2040/ viewed (Accessed 27 April 2021).

Fujita, M. (1989). '"It's All Mother's Fault": Childcare and the Socialization of Working Mothers in Japan'. *Journal of Japanese Studies*, 15 (1), pp. 67–91. doi: 10.2307/132408.

Government of Singapore. (2021). Reach, Singaporean Government Feedback Unit, *Conversations on Singapore Women's Development* [online]. Available at: www.reach.gov.sg/participate/conversations-on-singapore-womens-development (Accessed 20 April 2021).

Harding, R. (2021). Tokyo Olympic chief refuses to quit after saying women talk too much, FT, 4 February. Available at: www.ft.com/content/5b5b195a-d335-4784-b3cd-291e775f16f4 (Accessed 27 July 2021).

Heikkilä, M. (2020). 'Gender Equality Work in Preschools and Early Childhood Education Settings in the Nordic Countries – An Empirically Based Illustration'. *Palgrave Communications*, 6 (75). Available at: https://doi.org/10.1057/s41599-020-0459-7

Houseman, S., and Abraham, K (1993). 'Female Workers as a Buffer in the Japanese Economy'. *The American Economic Review*, 83 (2), pp. 45–51. Retrieved 5 May 2021, from www.jstor.org/stable/2117638

JILPT, Japan Institute for Labour Policy and Training. (2016). Available at: www.jil.go.jp/institute/research/2016/150.html (Accessed 25 April 2021).

Jones, R., and Seitani, H. (2019). 'Labour market reform in Japan to cope with a shrinking and ageing population', OECD Economics Department Working Papers, No. 1568, OECD Publishing, Paris. Available at: www.oecd-ilibrary.org/economics/labour-market-reform-in-japan-to-cope-with-a-shrinking-and-ageing-population_73665992-en (Accessed 26 April 2021).

Kakuchi, S. (2020). Record numbers of female students, but is it enough? *University World News*, 7 November. Available at: www.universityworldnews.com/post.php?story=20201 10608560151 (Accessed 25 April 2021).

Kano, A. (2015). The Future of Gender in Japan: Work/Life Balance and Relations between the Sexes. In Baldwin F. and Allison A. (Eds.), *Japan: The Precarious Future* (pp. 87–109). New York: NYU Press. Retrieved 6 May 2021, from www.jstor.org/stable/j.ctt15zc875.8

Kano, A. (2016). Labor of Love, Love of Labor: Debates on Motherhood and Work. In Kano, A. (Ed.), *Japanese Feminist Debates: A Century of Contention on Sex, Love, and Labor* (pp. 104–139). Honolulu: University of Hawai'i Press.

Kano, A. (2018). 'Womenomics and Acrobatics: Why Japanese Feminists Remain Sceptical about Feminist State Policy: Feminist Encounters', *A Journal of Critical Studies in Culture and Politics*, 2 (1), p. 06. https://doi.org/10.20897/femenc.201806

Learmonth. (2016). 'Consulta fined £2000 for discriminating against mother during interview'. *The National Scot News.* 31 May. Available at: www.thenational.scot/news/14903 272.consulate-fined-2000-for-discriminating-against-mother-during-job-interview/ (Accessed 21 March 2021).

Macnaughtan, H. (2015). 'Womenomics for Japan: Is the Abe Policy for Gendered Employment Viable in an Era of Precarity?' *The Asia-Pacific Journal*, 13 (12:1), pp. 4302

McCurry, J. (2015). 'Japanese politician in sexism row after call for women to have more babies'. *The Guardian*, 30 September. Available at: www.theguardian.com/world/2015/ sep/30/japanese-politician-yoshihide-suga-in-sexism-row-after-call-for-women-to-have-more-babies (Accessed 27 April 2021).

McCurry, J. (2018). 'Japanese politician in baby row thrown out again – for sucking cough drop'. *The Guardian*, 1 October. Available at: www.theguardian.com/world/2018/oct/ 01/japanese-politician-thrown-out-of-meeting-for-sucking-cough-drop-yuka-ogata-breastfeeding-row (Accessed 25 March 2021).

McKinsey & Co. (2021). 'Mobilizing women to step up as leaders in Japan', 27 April. Available at: www.mckinsey.com/featured-insights/asia-pacific/mobilizing-women-to-step-up-as-leaders-in-japan (Accessed 28 April 2021).

Mataharanet. Available at: http://mataharanet.org/en/what-is-matahara/ (Accessed 26 April 2021).

Matsui, K. et al. (2014). Japan: Portfolio Strategy Womenomics 4.0: Time to Walk the Talk, Goldman Sachs, 30 May 2014.

Matsui, K. et al. (2019). Japan: Portfolio Strategy Research Womenomics 5.0: 20 years on, Goldman Sachs, 18 April 2019.

Mukherjee, S. (2015). 'More educated and more equal? A comparative analysis of female education and employment in Japan, China and India'. *Gender and Education*, 27 (7), pp. 846–870. http://dx.doi.org/10.1080/09540253.2015.1103367.

Murphy, R. T. (2014). *Japan and the Shackles of the Past.* New York: Oxford University Press.

OECD.Stat. *Time spent in paid and unpaid work, by sex.* Available at: https://stats.oecd.org/ index.aspx?queryid=54757 (Accessed 20 April 2021).

Okazaki, M., and Johnson, G. (2013) *Kawaii!: Japan's Culture of Cute.* London: Prestel.

Osawa, M., and Kingston, J. (2015). Risk and Consequences: The Changing Japanese Employment Paradigm. In Baldwin F. and Allison A. (Eds.), *Japan: The Precarious Future* (pp. 58–86). NYU Press. Retrieved 6 May 2021, from www.jstor.org/stable/j.ctt15zc875.7

Parsons, S. (2017). Gender Parity: Reinvention of Business Culture and Norms, *Women in Leadership Publication*, Winter Issue, pp. 34–37 Available at: www.wilpublication.com/

Parsons, S. (2020). Is Gender Still on the Agenda in Japan? *East Asia Forum*. Available at: www. eastasiaforum.org/2020/11/11/is-gender-still-on-the-agenda-in-japan/

Rich, M. (2019). 'Two Men in Japan Dared to Take Paternity Leave. It Cost Them Dearly, They Say'. *NYTimes*, 12 Sept. Available at: www.nytimes.com/2019/09/12/world/asia/japan-paternity-leave.html (Accessed 16 April 2021).

Samadi F. (2016). 'Japanese Travel Agency Backs Down over College Girl Campaign' *PR Week* 12 May. Available at: www.prweek.com/article/1394628/japanese-travel-agency-backs-down-college-girl-campaign (Accessed 12 April 2021).

Shirahase, S. (2015). Demography as Destiny: Falling Birthrates and the Allure of a Blended Society. In Baldwin F. and Allison A (Eds.), *Japan: The Precarious Future* (pp. 11–35). NYU Press.

Song, J. (2018). 'Young people, precarious work, and the development of youth employment policies in Japan', *Japanese Journal of Political Science*. Cambridge University Press, 19 (3), pp. 444–460. doi: 10.1017/S1468109918000117

Steger, I. (2019) 'Why Japanese companies ban female staff from wearing glasses' *QZ*, 7 Nov. Available at: https://qz.com/1743901/why-japanese-companies-ban-women-from-wearing-glasses/ (Accessed 16 April 2021).

Street, Z. (2013). Fatherhood in Moral Education Textbooks. In Steger, B and Koch, A. (Eds.), *Manga Girl Seeks Herbivore Boy: Studying Japanese Gender at Cambridge* (pp. 83–128) Lit Verlag

Sugimoto, Y. (2010) *An Introduction to Japanese Society*. Cambridge, UK: Cambridge University Press.

Taylor M. (2021) 'Japan's asset managers under pressure to improve workforce gender diversity', *Institutional Asset Manager*, 2 February 2021. Available at: www.institutionalassetmanager.co.uk/2021/02/02/295327/japans-asset-managers-under-pressure-improve-workforce-gender-diversity (Accessed 24 April 2021).

Takami, T. (December 2018). 'Gender Segregation at Work in Japanese Companies: Focussing on Gender Disparities in Desire for Promotion'. *Japan Labor Issues*, 2 (11), pp. 7–12

Usui, C., Rose, S., and Kageyama, R. (2003). 'Women, Institutions, and Leadership in Japan'. *Asian Perspectives*, 27 (3), pp. 85–123. doi: 10.1353/apr.2003.0015

UN (United Nations) Department of Economic and Social Affairs, Sustainable Development, *The 17 Goals*, Available at: https://sdgs.un.org/goals (Accessed 3 April 2021).

WEF (World Economic Forum). (2021). *Global Gender Gap Report 2021*. Geneva: World Economic Forum, pp.8–9. Available at: www.weforum.org/reports/global-gender-gap-report-2021 (Accessed 16 January 2021).

Yamamoto, Y. (2016). 'Gender and Social Class Differences in Japanese Mothers' Beliefs About Children's Education and Socialisation'. *Gender and Education*, 28 (1), pp 72–88.

Yoshikawa, K., Kokubo, A., and Wu, C.-H. (2018). 'A Cultural Perspective on Gender Inequity in STEM: The Japanese Context'. *Industrial and Organizational Psychology*. Cambridge University Press, 11 (2), pp. 301–309. doi: 10.1017/iop.2018.19

Zhou, Y. (January 2021). 'How Women Bear the Brunt of COVID-19's Damages on Work'. *Japan Labor Issues*, 5 (28), pp. 2–8

12

GENDER AND SUSTAINABILITY IN OVACIK AND HOZAT COOPERATIVE ACTIVITIES, TUNCELI PROVINCE, TURKEY

İlknur Öner

12.1 Introduction

This study is based on fieldwork of two cooperatives in the frame of gender, sustainability, and culture triangulation from two townships (Hozat and Ovacık) of a Turkish province (Tunceli). It was important to explore and trace the ongoing ecological income-generating activity experiences to define, declare, and increase awareness, support, and applications for the future.

The chapter comprises three subheadings: introduction, discussion, and conclusion. The study examines the establishment structure of activities, the level of gender groups participation, the organizational structure of the gender groups, social relations, outcomes for the beneficiaries, training, and decision-making processes, transformation of structure and relations at the time of the research, benefits, and activity processes problems, power relations, social solidarity, and empowerment.

The historical evolvement of theoretical approaches to gender, development, and UN SDGs have created a multidimensional base for this study and it can be summarized as follows.

Previously, much research has focused on the promotion of gender equality, sustainability, income-generating activities, empowerment, development of resilient communities, and ecological approaches. First, the *meaning of sex, then gender issues* were important in the discussions. Before 1970, definitions of development and improvement relied on multiple factors. Thereafter, *power* and *participation* issues became important. Human resources started to be argued. The influence of *internal* and *external* multiple factors on improvement and development was differentiated (Barke and O'hare, 1984: 45–71). In time, economic development, poverty, and equal benefit were elaborated. Women's participation in economic development (Boserup, 1970), gender, and their participation (Momsen, 1991: 3) has gradually become the core study.

DOI: 10.4324/9781003174257-15

ILO's studies between the 1970s and 1980s systematically focused on women, development, and possible changes (Dixon-Mueller, 1985:i,v).

Furthermore, during the World Conferences on Women (Mexico City 1975, Copenhagen 1980; Nairobi 1985, Beijing 1995), women began to be recognized as a *heterogeneous* group and have *different activity lines*. Then, the *necessity of recognition and promotions* on gender equality, empowerment agenda, achievement, promotion, and follow-up in 12 areas were put forward in Beijing (UNWOMEN.org.2021).

UN established a commission (1946) to *monitor* the situation of women and promote women's rights. Declaration of the *Convention on Elimination of All Forms of Discrimination Against Women* CEDAW (18 December 1979 - New York) was important for the recognition of equality, discrimination, and rights and their *reachability* and *applicability in action for women* (ohchr.org.2021).

The Brundtland Report was important to understand the relationship between humankind and environment, scarcity, crisis, ecology, and sustainability by 2000 and beyond (World Commission on Environment and Development, 1987). The importance of the *needs, constraints,* and responsibilities of present and future generations were revealed by distinguishing differences of community and environments leading to the UN Rio conference (UNCED-1992, Rio) and the Local Agenda 21. Then, the *responsibilities of local governors* were labeled out framing their studies internationally, nationally, and locally for sustainability and development in the 21st century. Trade, environmental protection, lending issues between *developing and developed* countries, and *crises* became visible in discussions.

Despite this, gender issues discussions continued. The ethnicity, economy, and gender relations (Westwood and Bhachu, 1988) patriarchy and theories (Walby, 1990). Abadan-Unat et al. (1981) and Ozbay (1990) studied women's position, family, and social change in Turkey. In time, gender studies in Turkey included the following issues such as history, ecofeminism, politics of the body, manhood, Middle East, etc. (Doğu Batı, 2012–13). Gender role perspectives were studied sociologically and cross-culturally in the linkage of development, patterns, and possibilities (Lindsey, 1993; Spade and Valentine, 2011).

In time, methodologically gender issues and work, sustainability studies continued in some of follow-up studies (Katz and Monk, 1993; Oner, 2004).

Alternatively, some of the gender-paid work activities were based on general discussions (Gamarnikow et al., 1983; Game and Pringle, 1983; Pahl, 1988; Watson, 1993). However, some specific groups such as agriculture workers (Dixon-Mueller, 1985), women and development in the Third World (Momsen, 1991) and domestic workers, housework, housewife (Oakley, 1974), women's occupational mobility (Dex, 1987) were also explored. In Turkey, textile workers' experiences (Ecevit, 1986; White, 1994) and women cooperatives were studied (Varol, 2013).

Women and gender issues in studies, policies, and applications gradually occurred.

The impact of education, gaining income, increasing women's empowerment, and creation of change in their position were important points in gender issues (e.g. Deem, 1980; Abadan-Unat, 1981; White, 1994; Oner, 2004). Findings revealed that women's positions and power usage can *change* sustainability conditions of their activities either through their impetus, influence, or interference and mediation with other societal conditions. They were resulting *negatively*, *positively*, or *pausing* outcomes (Oner, 2004). Therefore, these quickly changing fragmented approaches, evaluations, and regulations relating to gender issues required continuity, follow-up, deep-down studies to support the promotion of women's bargaining power and empowerment (Abadan-Unat, Kandiyoti, Beauvoir). The importance of *empowerment, power transfer, life cycle* (Beauvoir, 1974), *empowerment* and women's *pseudo-emancipation* (Abadan-Unat, 1981), and *daily life* (Kandiyoti and Saktanber, 2003) *bargaining power of women and positioning of women in development policies* (Kandiyoti, 1988) was evaluated. In the 1990s, Oner's study of Keban Dam migrant females and the Turkish Sociological Association's work on GAP Project cases can be given as examples (Oner, 2004).

Alternatively, the question of *where* and *how* women were gaining income revealed many different arguments such as *integration* in the *dual labor market*. How much and in which sections' activities were *outlined* in *statistics*. Consumer patterns and usage of this generated income had become the focal point of *power and decision-making relations*-related theoretical discussions (Oner, 2004). *Working alone, in* groups, communities, or cooperatives may differentiate process, impact, recognition, and outcome. The expectation for women in the organized working group, in well-planned work processes, and in solidarity created a more protective environment in a complex and competitive labor market.

In time, the UN declared 17 Sustainable Development Goals for 2030. Through these ongoing initiatives supporting sustainable solutions with regional bodies'contributions, and benefits, economic growth built resilience and made it possible for them to attain more equitable relationships. Within the accumulated scope, like others, Turkish governing bodies (centrally and locally) began more promotion of such activities. Small-scale income-generating activities (such as knitting, crochet, lace, amigurumi; producing and selling natural products, etc.) were common ways for community growing projects to generate extra money to help fund their activities, livelihoods, reduce poverty, and increase equality and empowerment. Some of these activities could even include regional brand promotions development and management gaining domestic and internationally active sustainability of the brand. Studies were tracing these structures, organizations, and gender relations.

However, it remained unclear why men and women activity networks and the need for a combination of local, national, and international actors and interplays in existing or newly emerging social structures were important for the survival of cooperatives and the promotion of gender group activities.

Even when situations were stable before the COVID-19 pandemic, the following points had attracted: Strengthening capacities at the local level: strengthening the sustainability of initiatives, training delivery, participation; business development support, provisions policy note preparation on empowerment; prospective investors networking initiators, and participants; marketing support provision, collective activities, necessity for activision of gender groups, and international declarations signatories. However, in current living and working conditions, these were on a very slippery base in linkage to the frame of institutionalized relations, regulations, management, survivals, wealth, health, and safety nets in the current times.

The *purpose of this study* was to explore how two types of income-generating activities place gender groups in them. Moreover, how their intergroup dynamics and relationships within members and communities in specific environments shaped cultural group operations, and how the participants of such activities responded to the processes of such activities.

This study *methodologically* revolved around two empirical studies based on the participants of two cooperatives and two related brands (HOBE-Hozat Dolls) in Hozat and (Ovacık Natural Products) Ovacık Townships of Tunceli Province in Turkey, 2019.

HOBE-Hozat Doll is a brand of organic amigurumi knitting doll produced by local women's groups. Ovacık Natural Products brand labels agricultural products of a local cooperative. Thus, two types of income-generating activities—knitting and agricultural production—were studied.

The studies of these income-generating activities have a long history. Knitting had been at the core of discussions with feminism and feminity (Kelly, 2014), culture (Abrams, 2006), craft works, and studied in different countries.

ILO's rural work development and women studies in the 1970s focused on agricultural productions. Accumulated knowledge brought forward the importance of multidimensional factors, multidisciplinary complexity, and the need for collaborative activities in time. Recent examples are an internationally collaborative workshop on *Strategies for Economic Development in Southeastern Anatolia* which was held in Mardin, Turkey, in 2005 (UNDP, 2005). However, complexities focused on matters such as the *Sustainable and Just Rural Transitions: Connections and Complexities* IRSA XIVth World Congress of Rural Sociology in 2016. There were studies combining craft work, culture, and gender in the frame of new-old domesticity, digitalization, popular culture, political activism, and craft as power (Bratich and Brush, 2011).

There were international scrutinizations in the definitions as cooperatives and non-cooperatives, participation and leadership (Ozdemir, 2013; Bezbourah and Pilali, 2015; Dalkıran, 2017; Alarcon and Sato, 2019). The creation of brands following income-generating activities had been studied in the linkage to sustainability as well. "*Argande*" brand from Turkey was a recent example (Kipöz and Atalay, 2019).

The challenges and opportunities were important to bind gender and social-ecological resilience measures (Kawarazuka et al., 2017).

The *methodological frame* of this study was based on fieldwork and literature review. Technically, a questionnaire comprising 78 questions focused on both activities. Observations and semi-structured interviews (with participants, initiators, investors, local governors) were undertaken in April and December 2019 and onward.

Informants were not limited to gender groups. The focus was on *the cooperative members* and *cooperative activities*. Also, all of the relevant people to the income-generating activity processes were intended to be understood and included. This issue was important to understand *women's place in relation to others* and *their position and roles toward both gender groups in the cooperative processes.*

In Hozat, informants of the questionnaires from the Limited Responsible Hozat Entrepreneur Women Production and Management Cooperative (established the HOBE brand of Hozat Dolls) were biased on the women's side. This was based on knitting activity. Fifty participants were expected, but the eligible number was only 30. However, 29 of them were studied face to face in the activity room. The majority completed questionnaires themselves, and the researcher assisted illiterate informants.

The age range was between 22 and 68. Precisely, 20.6% were aged between 22 and 30, 17.2% between 31 and 41, 41% between 51 and 60, and 6.8% between 61 and 70. All informants were women including a responsible teacher. But male local governors and the mayor were also interviewed.

However, the type of income-generating activities of the Limited, Responsible Ovacık 94 District Agricultural Development Cooperative (established Ovacık Natural Products brand) required a complex and timely tracing process of informants.

At the time of field study, the allocation was male managers (3), producers (400), packaging (6), dairy section (2), trading and accounting, stitching bags (9), and sifting (23). An additional section in Tunceli was working with two in packaging and 5–6 in sifting (Ovacık Belediyesi, 2019).

The tracing process of some informants was problematic due to constraints on the type and phase of the activities, winter conditions, and small housing units. This was overcome by approaching them activity locations, houses, and understanding of household members. The expected number of cooperative members in Ovacık appeared in a large group (see activity structure section); available and eligible numbers of informants of the cooperative were 28 at the time of the study. Ovacık group activities included both gender groups. Most of the 28 questionnaires were completed by literate cooperative members. The researcher however assisted a few.

In both groups, illiterate informants were few and over 60 years.

At the time of fieldwork, tracing the process of women working in the chickpea and bean sifting plant was very time-consuming due to wintertime. Most of them had to be interviewed at home. Moreover, most producers were excluded (because of the need for further tracing in the villages and townships) at the preliminary field study, but some were included because it was possible to approach them due to their presence in Ovacık doing different activities for the cooperative.

The actual number appeared to be weighted in favor of women, but at the time of fieldwork, blurred division of members between first and second established cooperatives resulted in a sample group of 28 eligible cooperative members. The studied sample group of Ovacık included 66.6% female members and 33.3% male members. The age range of female members was between 25 and 60 and that of male members between 25 and 50. Females were overwhelmingly in the 41 and 50 age group but males were much younger between 25 and 30. Additionally, local governors were interviewed.

In both townships, aimed interview numbers were much higher, but this was impossible due to fieldwork conditions. Interviewing was much easier to reach in Hozat because they were all together. However, convincing informants in Ovacık was more difficult because informants were widespread and have confidentiality issues and impact of terror in the province. This felt like the efforts of the citizens of townships to contribute to their future image but existing contradicting feelings were negligible which was overcome by necessary explanations.

Despite much more collaborative and conscious approaches, the informants of both township cooperatives refused to accept type recordings or signing consent of a questionnaire form. This was due to a breach of promise confidentiality rules whereby it was possible to trace their responses.

Preliminary and limited qualitative findings were evaluated in this chapter. Gender, the establishment of activities, organization, participation, the meaning of activities, social relations, training process, decision-making, power relations, solidarity, empowerment, transformations, benefit, and problems. Overall findings were analyzed in more detail through SPSSX and MAXODA but limited definitive summaries were used here.

Findings revealed that although they were not mentioned, UN SDG goals were traceable and interconnected in the observed processes of two cooperatives. 2 - Zero hunger, 5 - Gender equality,10 - Reduce inequality within and among countries, 11 - Sustainable cities and communities, 12 - Responsible consumption and production, 13 - Climate action, 15 - Life on land, 16 - Peace, justice, and strong institutions, 17 - Partnership. These findings were summarized in the following section.

12.2 Discussion

Preliminary findings of the fieldwork in two townships are summarized under ten subtitles. This section covers findings on the structure of activities, organizational structure, participation, the meaning of activities, social relations, decision-making processes, transformations, benefits, and activity problems.

12.2.1 Establishment structure of activities

Dixon-Mueller (1985), , Hakim (1988), Watson (1993) and Oner (2004) revealed that recognition and visibility of women's work differ depending on time, casualty, regularity, and possible orientations.

Observations on cooperative activities revealed that paid work and monthly or daily income-generating activities did not guarantee the works of some active members (SDG 16). This was related to the type, irregularity, and phases of seasonal work (SDGs 13 and 15). For example, doll production in Hozat could be done at the official room or home and even if they are away from the town. However, this was not possible for some cooperative members in Ovacık because of the production phase, seasonal activity, and daily/seasonal establishment structures. Although the enlarged listing of agricultural producers (including landowners and workers) produced a general and slight regularity of daily workgroups which created irregularity (SDGs 5 and 10). This comune-based work structure of Ovacık was established with the initiation of the former mayor called "communist president" (now mayor of Tunceli city) had been taken as a model for some other cooperatives in Turkey. They combined and used possible production fields (state and private) (SDGs 10 and 11).

The related findings on the structure of paid work approached two levels and their interconnections (cooperative and informants' core experiences).

The first was the *type of paid work* activities that formed the basis for *cooperatives.* This chapter focused mainly on the output of those structure-related findings (see Figure 12.1).

The second was the frame of the *experienced paid work* activities of *informants* (see Table 12.1).

Detailed work activities of individuals were traced concerning monthly distributions and in/out paid work activities with drilling questions which are not discussed here.

<table>
<tr><td rowspan="4">OVACIK</td><td>Heads of cooperative 3 males</td><td colspan="2">400 agricultural producers (excluded at preliminary stage of the study) This group includes locals and then extended to some producers of other townships of Tunceli in time</td><td>4 male watchmen of common production field in Kızık village, four months in Summer.</td></tr>
<tr><td>Stitching cloth packaging bags-9 insured members</td><td>Sifting plant (alternately work and daily wage) 6 male 80 female. At the time of reachable number was 23</td><td>Irrigation, harvesting, patos-seperation machine, transfers at allocated field in Kızık village 10 people</td><td>Dairy section 2 people (female and male)</td></tr>
<tr><td colspan="2">Trading, accounting, posting 4 people from both gender groups+ insured</td><td colspan="2">Tunceli city section: 5-6 people in sifting and 2 people in packaging</td></tr>
<tr><td colspan="4">Trading and marketting
Digital platforms and five actual shops in Turkey (numbers are increasing). local governors, mayor, locals and national and international consumers</td></tr>
<tr><td rowspan="4">HOZAT</td><td colspan="4">Cooperative members (10 core members: 8 members from Hozat and Local Governorship respresentative Female who is working at Tunceli ; Responsible teacher-Female and was in Hozat then left for Izmir. Other 40 producers of the cooperative</td></tr>
<tr><td colspan="4">Material provision
(all gender groups, locals, governors, mayors, businesman and business women and cooperative members)</td></tr>
<tr><td colspan="2">Mainly Knitting- HOBE doll, scarf and berets by female members</td><td colspan="2">Rarely Woving (cicim, handbags, wallets, rugs and others) by female members</td></tr>
<tr><td colspan="4">Trading and marketting
Cooperative members, the head of cooperative and reasponsible teacher, local governors, mayor, locals and national and international consumers. They are connected actual and digital platforms. They initially used Ovacık's marketing shops in Turkey but they were on the way to establish their own brand) Attained local and national fests</td></tr>
<tr><td></td><td colspan="4">CONSUMERS OF BOTH COOPERATIVE PRODUCTS
(Local, national and international)</td></tr>
</table>

FIGURE 12.1 Structures of studied Hozat and Ovacık cooperative activities.

Source: Figure generated from information of Ovacık Belediyesi (Ovacık Municpiality); 2019 and Hozat Kaymakamlığı (Hozat Governorship); 2018, 2019.

TABLE 12.1 The frame of individual forms of activities of two cooperative members

Farm-production
Agricultural works (first, second, third product; provision of land, clearing land, preparing the land, provision of seeds, organizing workers, sowing/planting, irrigation, fertilization, weeding, pest control, harvest, threshing, product processing/fabrication, marketing)
Animal rearing (livestock) (first, second, third type of animal; provision of animal, organizing workers, shepherding (herding), feeding, milking, product processing, marketing)
Home gardening (vegetables, fruit trees, small animals/poultry)
Non-farm production
Handicrafts (products, provision of material, production, control, marketing)
Home industry (Tomato/paper paste, jam, dried vegetables, walnut, others)
Trading, purchase-sale
Services (child-rearing, elderly care, cleaning, etc.)

Source: Developed from Dixon-Mueller's classification (1985:125–135).

This overall approach revealed much more complex networks and patterns, but initial findings were limited to a preliminary summary of the first phase (the type of paid work activities that formed the basis for cooperatives).

HOBE doll producing cooperative activities, Hozat knitted dolls project established by Hozat Local Governorship in 18.03.2018 (Hozat Kaymakamlığı, 2018). Production of HOBE dolls covered several activities by the time of pandemic. SDGs 2, 5, 13, 15, 10, and 16 related initiations were important. These activities can be summarized as follows:

Provision of eco-friendly materials (SDGs 13 and 12): organic yarn, cotton, pattern provided by the local governor, local business people, cooperative members, and vocational teacher. They were provided from outside of the township and sometimes other provinces (except patterns).

The organization of working groups and labor division was provided by a vocational teacher and the collaboration of cooperative members.

A specific official room and informants' homes were locations for knitting. Labor division in the activity room depended on the type of production (knitting, filling model, stitching, labeling, pricing) and provision of refreshments (daily selection of serving persons in turn). Although, the HOBE dolls brand included traditional, local Cicim weaving (handwoven rug modified as a handbag, pillowcase, backpack, wallet, saddlebag, prayer rug, car seat cover), and scarf and beret knitting as well. Their production was rare (SDGs 2, 5, 12, and 13).

This cooperative comprised production activity and administration. The trading was mainly carried out by local governors, the teacher, and networks of local members. They utilized Facebook, Twitter, and Instagram accounts, but mainly one person a teacher carried out these activities (as the president was otherwise engaged).

These cooperative members initially were public training center attendees of Hozat. Subsequently, the governor of Hozat, with the support of the governor of the province initiated them and financial support was gained. First, the cooperative president was elected on 30 June 2018 as a local government representative. Despite this, this cooperative was declared the first women's cooperative in Turkey, with the public institution (Local Government) as the founding partner. However, the established process and activities were different in Ovacık.

Ovacık-*the Ovacık Doğal* brand was the outcome of efforts of the Limited Responsible Ovacık 94 District Agricultural Development Cooperative initiated by Ovacık Municipality in 2017. They produced ecological honey and honey products, natural salt, mulberry and boiled mulberry /grape juice, agricultural product, legumes, and dairy products (SDGs 12, 13, and 15). The president and vice president worked voluntarily. Ovacık municipality was a pioneer for the establishment, but it was not a partner of the cooperative. Ovacık and Hozat cooperatives were different in production, structure, and management (Ovacık Belediyesi, 2019: 99).

The organization structure fell within the frames of SDGs 2, 12, 15, and 10, but the institutions' strength was on different levels in both cooperatives concerning SDG 16. Ovacık was relying mainly on the local community and Hozat was mainly dependent on the leadership of the local governorship leadership.

12.2.2 Organizational structure of the gender groups

According to SDGs 5, 16, and 17, weights of gender groups in the activities in both sections differed as follows:

Hozat—The establishment of the cooperative was based on female members; however, some men were active in the processes except knitting. They appeared as raw materials financial providers, local government managers, customers, and household social supporters.

Ovacık—The activities lacked gender bias formally. However, there were some traditional gender workload divisions during different production and marketing processes. These were complicated, and it is impossible to detail them here. The aim was to create a communal work structure for both women and men. Management and trading relied mainly on men, but patterns changed according to products and work processes.

Locations of the gender groups in Hozat were more clear-cut than Ovacık. However, Ovacık had more flexibility to act together in production, marketing, and support.

12.2.3 The level of gender groups participation

In Hozat, women mainly ran knitting activities. However, raw materials provision and trade required men's help because of their financial power, acceptability as initiators to search linkages and digital abilities.

Ovacık: Agricultural production types defined the activity of gender groups. Field plowing and irrigation were accepted as men's work. Sowing crops were done by both gender groups. Hoeing, picking beans, and chickpeas were mainly carried by women. Buying seeds and selling products were usually men's work if they bought at cooperative stores. However, sales were done by a responsible person irrespective of gender at the cooperative store.

Thus, gender equality applications appeared differently in both participating cooperatives.

12.2.4 Meanings of activities for beneficiaries

The meanings of activities for beneficiaries were important and can be linked to SDGs. They were important for the continuity, survival of informants and their households (SDG 2), and the strength of the institutions, cooperatives, (SDG 16), and their connections were able to create aimed SDGs 5, 10, 11, and 12's conditions. There were different activity beneficiaries, but initial findings can be revealed as follows:

Hozat—The formal meaning of activities for the *local governor and initiator*: In the economically underdeveloped region, the aim was to increase women's economical income by providing their participation in the labor force. Divorces at a young age were accepted high in the township at that time. Women's contribution to household income was intended to strengthen family unit (Hozat Kaymakamlığı, 2018). SDGs 5 and 10 were related goals. However, women were treated as supplementary income generators in this frame of Hozat initiation.

Ovacık—The meaning of activities as *municipality and initiator at the formal frame*: These were essential for collaboration and bringing together production (Ovacık Belediyesi, 2019: 99). SDGs 12, 16, and 17 were in the front line, and there were no segregated gender indications that could be considered with the equality approach.

Hozat and Ovacık—The meaning of activities for the *members* were complicated and had different meanings according to priorities at the time which are out of the frame of this section.

12.2.5 Social relations

Hozat-Social relations among the cooperative members were closely linked. However, there was competition among members about product selection and sales. Hozat community supported the activities of the cooperative. Product marketing was institutionalized with the local governorship and traditional local fellow-citizenship network lines (nationally and internationally). Social relations between local governors and cooperative members were essential, but the local governors' attitudes toward income-generating activity types directly affected women cooperatives. The HOBE group lost their activity room because of the new local governor's attitude and the influence of the pandemic lockdowns. The previous local governor was a supporter and establisher initiator of the activity. However, the new local governor was more interested in agricultural production and other types of income-generating activities.

Ovacık-Social relations between cooperative members are closely linked. There was not much competition at the production level in appearance. Producers knew that their products will be taken by the cooperative. Although daily groups regularly called, in turn, to work, there was sometimes competition for the establishment of the daily women working group members. Despite this, they helped each other during the production processes. Product marketing was more clear-cut in Ovacık because Ovacık cooperative had its distribution centers at local and national levels. They were successfully using fellow-citizenship networks and digital platforms for marketing. For example, the snowball effect to reach new consumers was used in both cooperatives.

Linkages between cooperative and local governors were essential for Ovacık as well (SDGs 16 and 17 are closely linked). However, conflict arose between local governors and cooperative members, which was the case for the storage building and its problems. Different political approaches by local governors also had an important influence on the sustainability of cooperative activities like in Hozat. This was important for partnership constraints.

12.2.6 Training processes

Essential training processes were necessary for the survival of both cooperatives (SDG 16). However, recently important training courses such as computer usage, marketing, and managerial skills were mainly done by Ovacık cooperative members. Whereas Hozat appeared to be trained for handicraft. They lack much-needed skills such as digitalization, marketing, and management. The migration of the only responsible teacher to Izmir without a skilled replacement put a constraint on the survival of the cooperative. This confinmed the importance of the connection of sustainability studies, educational attainment, and inequality studies (Jencks, 1972). There was an urgent need for formally recognizable skilled people. In application, skilled members did not poses the necessary qualifications which were difficult to achieve because there were constraints. Cooperative survival and sustainability due to the pandemic and different visions of recent local governors and lack of institutionalization of activities (SDGs 16 and 17).

12.2.7 Decision-making processes

Decision-making processes were considered (household; marriage, education; finding work; inside work organization) and could be linked to SDGs 5, 15, 16, and 17. Participation in such decision-making processes, the final decision, and the ability to change the result was important. Certain decision fields were linked with people's responsibility levels, which were *decisions related to production* and *household production, and reproduction*. Decisions relating to *production* were divided into agriculture, income-generating activity, and paid work. *Agricultural activities* included labor distribution, planting products, input and tool supply, hired workers supply, participation in a cooperative or loan program, sales, income control, and consumption.

Income-generating activities included decisions linked to labor distribution, activities selection, input and tool supply, income control, and consumption. *Paid work* activities included decision-making regarding the entrance to work, income control, and consumption. *Household management and reproduction* decisions included consumption, household management, and reproduction. *Consumption* included food, clothing, home maintenance/repair, home supplies, children's schooling, medical treatment, events, vacation time and place, leisure activities. *Household management and reproduction* have consisted of the division of labor, food distribution among household members, child guidance, family sizes, and others. These were extended with other drilling questions. Detailed findings are excluded from this chapter.

12.2.7.1 Power relations

Power relations were included in the above and other decision-making fields. Findings were gathered with cross-checking and drilling questions on education, marriage, leaving the job, social solidarity, etc.

12.2.7.2 Social solidarity

Social solidarity was much closer and more powerful in Ovacık agricultural production in the group (SDG 11). Observations revealed that members relied on the cooperative and their environmentally friendly organic products. Locals of townships, provinces, and other cities and abroad created important online network support, thereby eliminated transportation costs. Hozat had similar support from local citizens despite being less organized digitally from the Ovacık case. This apart, the work type of Hozat was more homogenous and there was solidarity to teach one another craftwork.

The cooperatives of Hozat and Ovacık were also in solidarity. With the influence of Ovacık's agricultural activities, its media popularity created an emerging agricultural cooperative with additions to the doll production. This increased some women's workloads all year around. Moreover, HOBE products, Hozat Dolls were sold in some of Ovacık's organic product centers in other cities. In 2018, there were five centers (Ovacık Belediyesi, 2019: 99), but numbers have been increasing. These social solidarity activities in relation to cooperatives' activities can also be observed at the cooperative fairs and townships, and the provincial festivals. And they could become much more powerful with the support of local and central governors. Other towns were in solidarity with these cooperative activities such as the initial production of cloth bags for the Ovacık Natural brand. SDGs 17, 16, 12, and 11 are at different levels in both cooperatives.

12.2.7.3 Empowerment

Empowerment will be the focus of another study. Generally, women and men in both townships relied on and were attached to their activities. However, the empowerment of the Ovacık cooperative's members was observed during interviews. The structure of commune-type organizations and members' meetings had given a strong sense of belonging and activity related to SDG 11. Therefore, SDG 5, gender equity, was important for both cooperative members.

The Hozat and Ovacık studies revealed that working in cooperatives strengthened positions. However, the guarantee that products will be taken by the cooperative gave producers a safety net in Ovacık. Such safety did not provide the same strength in daily working group members of the fieldworkers who were selected randomly. Similar hesitance could be observed in Hozat. Some producers were more skilled and their products could be sold easily. Cooperative applications revealed different outcomes regarding SDG 2-zero hunger. Product quality empowered some members more because of higher income. There was no such segregation in the selection of the Ovacık products but some existed at the establishment of hoeing and harvesting daily groups. These relations affected members' power relations and empowerment differently.

12.2.8 Transformation of structure and relations in time

There were great transformations at the time of cooperative activities. In the Hozat case, informal activities became formal cooperative and brands. In the Ovacık case, activities popularized and reached more consumers quickly with more markets. They became pioneering cooperative activities. Both cases created their ecologically friendly organic brands (SDGs12 and 13). However, organic products were costly and difficult to sustain. The support of local governors and other networks diminishing (SDG 17-Partnership) putting constraints on cooperative activities and strength of the institutions (SDG 16 peace, justice, and strong institutions). Both products were introduced in mass media. This was valuable but could not gain aimed strength and support. Ovacık had been preparing to include handcrafts such as knitting dolls, bringing the attempt in the frame of SDG 11 sustainable cities and communities to the forefront.

12.2.9 Benefits and problems of the activity processes

The benefits of activities differed. They generated income for members, introduced two forgotten and lonely, mountainous townships which suffered long from terror (SDG 8-decent work and economic growth, 10-reduce inequality, 13-climate action, and 16-peace, justice, and strong institution). They were introduced touristically with nature, hospitality, and eco-friendly products. Activity processes had difficulties. Hozat dolls could be produced in more livable conditions. However, the production of Ovacık agricultural products required working flexibility and collaboration. Because of climatic conditions packaging and storage in large areas without proper heating, and open-air crop processes. For example, if products were selected at home, it was easy to carry on. If a similar process were done in a big storage section, difficulties more in winter. Solidarity was incredible, although some competition existed between cooperative members and other cooperatives. Such experiences emphasized by Kışlalıoğlu and Berkes as revealing the importance of nature and human collaboration (2003).

Hozat products were sold per piece. Ovacık cooperative signed agricultural production contracts with producers whereby production costs were transferred to the bank accounts of the producers. Daily working groups received daily wages. Products were sold by weight.

Some products of both cooperatives were distributed by governors as gifts and advertisements of the town.

12.3 Conclusions

The study elaborated preliminary findings of the Limited Responsible Hozat Entrepreneur Women Production and Management Cooperative-(HOBE-Hozat Dolls) activities and the Limited Responsible Ovacık 94 District Agricultural Development Cooperative (*Ovacık Doğal* -Ovacık Natural Products).

The findings indicate that most countries will be concerned with UN SDGs by 2030. The selected cooperatives of two townships of Tunceli province exemplify that cooperatives could operate under different circumstances and outcomes in the embedded social texture. Appearances are affected by the collaboration of many factors such as culture, society, politics, economy, and health.

However, members of these activities believe in them which changes their narrow, medium, and large societal environments. These changes occur through their participation in eco-friendly, organic productions. Their living conditions are affected, strengthened, and resiliencies are improving but there are constraints. While such cooperatives have challenging difficulties, they continue to affect consumers and societies with their organic and eco-friendly products. The COVID-19 pandemic highlighted the importance of understanding nature and humans together. The gender bases may differ in cooperatives, but they require strong support. Sustainability is very important and has to be approached multidisciplinary. Agricultural activities, animal rearings, organic, eco-friendly products, the possibility of famine, new societal order, etc. are recent discussion topics. The two cooperatives of Tunceli province have pioneering efforts. They are closely related to several UN SDGs (2, 5, 10, 11, 12, 13, 15, 16, and 17) locally accumulated and activated cultures. Turkey is initiating to combine all cooperatives in a database. However, current constraining conditions (like in the Hozat case which is putting initiation on to the edge of closing down) require open-mindedness, collaborative actions, local, national, and international approaches, and analysis of present conditions to create support for a better future and survival of our communities.

References

Abadan-Unat, N., Kandiyoti, D., and Belik-Kıray, M. (1981) *Women in Turkish Society*, Leiden: E. J. Brill.

Abrams, L. (2006) Knitting, autonomy, and identity: The role of hand knitting in the construction of women's sense of self in an island community, Shetland, c.1850–2000. *Textile History*, 37(2): 149–165.

Alarcon, J.M.S., and Sato, C. (2019) Enacting peasant moral community economies for sustainable livelihoods: A case of women-led cooperatives in rural Mexico. *World Development*, 115: 120–131.

Barke, M., and O'hare, G. (1984) *The Third World*, Edinborough: Oliver & Boyd.

Beauvoir, S.D. (1974) *The Second Sex*, London: Penguin Books.

Bezbourah, K.C., and Pillai, V. (2015) Exploring the participation of women in financial cooperatives and credit unions in developing countries, *Voluntas*, 26: 913–940.

Boserup, E. (1970) *Woman's Role in Economic Development*, London: George Allen & Unwin.

Bratich, J.Z., and Brush, M. (2011) Fabricating activism: Craft-work, popular culture, gender. *Utopian Studies*, 22(2): 233–260.

Buergenthall, T. (1991) *International Human Rights*, Minnesota: West Publishing Company.

Dalkıran, G.B. (2017) The support of women work within cooperative enterprises: Sample of Turkey. *Social Research Journal*, 6(3): 1–11.

Deem, R. (ed.) (1980) *Schooling For Women's Work*, London: Routledge.

Dex, S. (1987) *Women's Occupational Mobility, Lifetime Perspective*, UK: Palgrave Macmillian.

Dixon-Mueller, R. (1985) *Women's Work in Third World Agriculture*, Geneva: International Labour Office, Women, Work, and Development Series No:9.

Doğanay, F. (1997) *Sosyal ve Kültürel Dönüşme Sürecinde Harran Ovası ve GAP Uygulaması*, Ankara: *DPT:* Sosyal Sektörler Genel Müdürlüğü.

Doğu Batı. (2012 13) *Doğu Batı, Toplumsal Cinsiyet*, Kasım, Aralık, Ocak, No:63.

Ecevit, F.Y. (1986) *Gender and Wage Work: A Case Study of Turkish Women in Manufacturing Industry*, Ph.D. Thesis, UK: University at Kent and Canterbury.

Game, A., and Pringle, R. (1983) *Gender At Work*, London: Pluto Press.

Gamarnikow, E., Morgan, D., Purvis, J., and Taylorson, D. (eds.) (1983) *Gender, Class & Work*, England: Gower.

Hakim, C. (1988) Homeworking in Britain, In. Pahl, R.E. (ed.) *On Work, Historical, Comparative & Theoretical Approaches*, Oxford: Basil Blackwell, pp: 609–632.

Hozat Belediyesi. (2019) *Hizmet Bülteni* (Buletin of Services), Hozat: Hozat Belediyesi (Hozat Municpality).

Hozat Kaymakamlığı. (2018) Hozat bebekleri projesi, Hozat Kaymakamlığı, www.hozat.gov.tr/hozat-orgu-bebekleri (Date of Acces: January 2019).

Hozat Kaynmakamlığı. (2019) *S.S. Hozat Girişimci Kadınlar Üretim ve İşletme Kooperatifi Hozat Bebekleri* (S.S. Hozat Entrepreneur Women Production and Management Cooperative Hozat Dolls), Hozat: Hozat Kaymakamlığı (District Governor of Hozat) (Date of Acces: January 2019).

IRSA (2016) *Sustainable and Just Rural Transitions: Connections and Complexities, IRSA XIVth World Congress of Rural Sociology*, 10–14 August. www.ryerson.ca/arts/irsacongress2016/call-for-submissions/call-for-sessions.html (Date of Access: April 2020).

ILO (2013) *Trade Unions and Cooperatives: Where Are We At?* Geneva: ILO Series: Journal of International Labour Research v. Fifth Edition eBook.

Jencks, C. et al. (1972) *Inequality; A Reassessment of the Effect of Family and Schooling in America*, Harmondsworth, Middlesex: Penguin Books.

Kandiyoti, D. (1988) Bargaining with Patriarchy. *Gender & Society*, 2(3):274–290. https://doi.org/10.1177/089124388002003004 (Date of Access: January 2021).

Kandiyoti, D., and Saktanber, A. (2003) *Kültür Fragmanları, Türkiye'de Gündelik Hayat* (Fragments of Culture, The Daily Life of Turkey) Istanbul: Metis Yayınları.

Katz, C. and Monk, J. (eds.) (1993) *Full Circles Geographies of Women Over the Life Course*, London: Routledge.

Kawarazuka, N. et al. (2017) Bringing analysis of gender and social-ecological resilience together in small-scale fisheries research: Challenges and opportunities. *Ambio*, 46: 201–213.

Kelly, M. (2014) Knitting as a feminist project? *Women's Studies International Forum*, 44: 133–144.

Kışlalıoğlu, M., and Berkes, F. (2003) *Ekoloji ve Çevre Bilimleri*, İstanbul: Remzi Kitabevi, Dördüncü Basım.

Kipöz, Ş., and Atalay, D. (2019) Sustainable Anatolia: Craft-Centered Design, In. Gwilt, A., Payne, A., and Ruthschilling, E.A., *Global Perspectives on Sustainable Fashion*, USA: Bloomsbury Visual Arts, pp:164–173. DOI:10.5040/9781350058170.ch-004.5 (Date of Access: April 2021).

Lindsey, L.L. (1993) *Gender Roles, A Sociological Perspective*, New Jersey: Prentice-Hall. Second Edition.

Momsen, J.H. (1991) *Women and Development in the Third World*, London: Routledge.

Oakley, A. (1974) *Housewife, High Value-Low Cost*, London: Penguin Books.

OHCHR (2021) "Convention on the elimination of all forms of discrimination against women New York, 18 December 1979," *United Nations Human Rights Office of The High Commissioner*, www.ohchr.org/en/professionalinterest/pages/cedaw.aspx (Date of Access: January 2021).

Oner, I. (2004) Kadınların Kırdan Kente Yeniden Yerleşime Cevabı: Elazığ'da Keban Barajı Kadın Göçmenleri Örneği, *Unpublished Ph.D. Thesis*, Elazığ: Fırat University, Institute of Social Sciences, Department of Sociology.

Ovacık, B. (2019) *Ovacık 1 Nisan 2014- 1 Ocak 2019 Yerel Yönetimler Programı (Ovacık 1 April 2014-1 January 2019 Local Governors' Programme)*, İstanbul: Sancı Yayınları.

Ozbay, F. (ed.) (1990) *Women, Family and Social Change in Turkey*, UNESCO.

Ozdemir, G. (2013) Women's cooperatives in Turkey 1st World Congress of Administrative & Political Sciences, *Procedia-Social and Behavioral Sciences*, 81: 300–305.

Pahl, R.E. (ed.) (1988) *On Work, Historical, Comparative & Theoretical Approaches*, Oxford: Basil Blackwell.

Spade, J.Z. and Valentine, C.G. (2011) *The Kaleidoscope of Gender, Prisms, Patterns, and Possibilities*, Thousand Oaks, CA: Pine Forge, Third edition.

UN WOMEN (2021) World Conferences on Women *UN WOMEN*. www.unwomen.org/en/how-we-work/intergovernmental-support/world-conferences-on-women (Date of Access: January 2021).

UNDP (2005) *Strategies for Economic Development of Women in Southeastern Anatolia workshop/brainstorming*. www.undp.org/content/dam/turkey/docs/news-articles/UNDP-TR-MardinWorkshop%Sept-29th-30th-2005%20Program.pdf (Date of Access: April 2020).

Varol, F.C. (2013) An Analysis of Women's Cooperatives in Urban Turkey in Terms of Women's Empowerment, *Unpublished M.A Dissertation*, Ankara: Middle East Technical University, Institute of Social Sciences, Dept of Sociology.

Walby, S. (1990) *Theorizing Patriarchy*, Oxford: Blackwell Publishers Ltd.

Watson, T.J. (1993) *Sociology, Work & Industry*, London: Routledge.

Westwood, S., and Bhachu, P. (eds.) (1988) *Enterprising Women: Ethnicity & Economy, Gender Relations*, London: Routledge.

White, J.B. (1994) *Money Makes Us Relatives, Women's Labor in Urban Turkey*, Austin: University of Texas Press.

World Commission on Environment and Development. (1987) *Our Common Future*, Oxford: Oxford University Press. https://sustainabledevelopment.un.org/content/documents/5987our-common-future.pdf (Date of Access: January 2020).

INDEX

For Product Safety Concerns and Information please contact our EU
representative GPSR@taylorandfrancis.com
Taylor & Francis Verlag GmbH, Kaufingerstraße 24, 80331 München, Germany

www.ingramcontent.com/pod-product-compliance
Ingram Content Group UK Ltd.
Pitfield, Milton Keynes, MK11 3LW, UK
UKHW020005200926
475702UK00015B/2485

* 9 7 8 1 0 3 2 0 0 4 5 2 5 *